AF346284

Silicon Carbide Revolutionizes Electric Vehicles: Unlocking High-Power Applications

Class

Copyright © 2024 by Class

All rights reserved. No part of this book may be reproduced in any manner whatsoever without written permission except in the case of brief quotations embodied in critical articles and reviews.
First Printing, 2024

Table of Contents

1 Introduction

The consumer demand for wide-band gap Silicon carbide semiconductors has been accelerated exponentially due to their superior material performance and high efficiency. The WBG industrial growth is expected to increase from current figure of 750 million € to 1812 million € at the end of 2025 [1] .

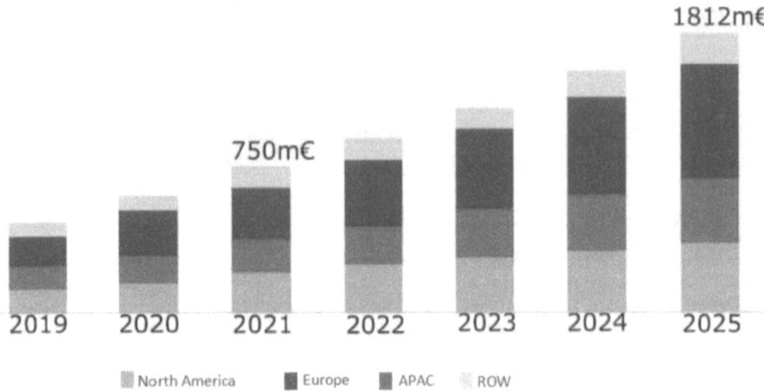

Figure 1 Silicon Carbide market growth between 2019-2025 (EURO Millions)

1.1 Background and problem motivation

SiC technology enabled electric vehicles to meet the design parameters required in high-power applications leading to better system performance and longer reliability. Global industries continuously make an effort to enhance the performance of available SiC devices. An increasing number of SiC MOSFETs with voltage ratings between 600 V and 1700 V are hitting the market with decreasing Rds (on) and costs as well as sufficient reliability. We wanted to make better the performance of available SiC devices with suitable application and system environments. The simple and traditional method for driving the MOSFET is the direct drive from the logic signal. Such method has a low current drive and slower transition with lower gate to source (V_{GS}) magnitude. The microcontroller usually provides near 20mA current which is not substantial for turn on high density power MOSFETs. The lower V_{GS} around 5V or less cause issues with the MOSFET that requires 12 to 20V to achieve the best performance. Some logic level power MOSFETs exist but high power SiC MOSFETs needs a faster and reliable gate driver approach to match their performances [2]. This imposed the demand to have a MOSFET with extremely low resistance and a gate drive with low inductance design. Additionally, on the driver side highly integrated gate drivers with suitable protection features, low inductive paths, and optimized driving capabilities are required for higher frequency applications to minimize the overall losses [3].

1.2   Overall aim

The overall aim of this book is to develop and optimize the gate driver solutions for SiC MOSFETs for the transport and industrial applications. The gate driver solutions should address all the issues related to noise in the system, overvoltage, and short circuit protection features. The designed SiC gate driver should reduce the spikes and rigging under normal and short circuit conditions to enable the dramatic increase in power density.

1.3   Concrete and variable goals

The work is emphasised on achieving high electrical efficiency from SiC switches and their gate drivers in advanced switch-mode-power-supplies (SMPS) to align it with commercial applications. The study-specific objects were:

1. Developing and optimizing a gate driver solution for a SiC MOSFET for achieving high power density.

2. Comparing the influence of different gate drive modules on SiC switches

3. Evaluation of feasible and commercially available SiC MOSFETS and further characterization on double-pulse test setup.

4. Investigating the switching transitions of the SiC switches and gate drivers were thoroughly.

5. Building-up a prototype with DC-DC converter with advanced SiC switches and gate drivers.

6. Investigating the impact of latest gate drivers and SiC switches in DC-DC converters. In the end, SiC based DC- DC converter was optimized with suitable protection features.

7. Determining the electromagnetic interference (EMI) due to fast switching transitions.

1.4 Scope

The work done in book led to promising results regarding the use of advanced gate drivers in power converters. The designed gate drivers address the critical challenges that arise in operating SiC MOSFETs at high voltage and frequencies. The designed driver modules effectively help in reducing the EMI and switching losses. The designed gate drivers offer an ultimate solution to the industry, which is thriving to attain energy efficient systems as per new EU regulations.

1.5 Outline

Chapter 2 provides a background overview of SiC MOSFETs, their basic structure, comparisons between Si and SiC MOSFETS, and IV characteristics. It also includes the switching characteristics and influence of gate drivers on SiC MOSFETs.

The potential applications of SiC transistors such as clamp inductive load, switching loses, and gate driver techniques are given in chapter 3. This chapter presents a general background of the SiC MOSFETs in the switch mode applications. The first section covers the theory related to analyse switching transients and calculate the power losses in the applications. While section two shows the background about SiC MOSFETs based half-bridge DC-DC converter and gate driver design approach.

Chapter 4 shows the implementation and design of the gate driver modules, different stages of design, power supply design, PCB layout and the selection of gate driver IC. The chapter 5 explains the design for test approach such as double pulse test description, measurement setup and methods.

Chapter 6 explains the results of double pulse test, and the selection of SiC switches and Schottky diodes including the influence of gate resistance, dV/dt, and dI/dt within the switching circuit. It further explains the driver strength, DPT switching loses, propagation delay, and short circuit protection test. Simulation verification and experimental analysis are also described in the chapter.

Chapter 7 provides a background of prototype converter, and the designed parameters. Chapter 8 covers the experimental validation of prototype converter and presents the results of these gate drivers and SiCs. Multiple tests have been conducted on various configuration to successfully demonstrate the functionality of prototype converter with these advancements are shown.

1.6   Contributions

The project was conducted mainly at STC research centre, Sundsvall, Sweden, in active collaboration with industrial partners, PowerBox AB, Gnesta and RISE Acreo, Kista, which provided essential insights, useful resources, and access to the latest technologies. The project was funded by Swedish Energy Agency and PowerBox AB.

2　Theory / Related work

SiC MOSFETs are wide-band gap devices consisting better properties as compared to Si semiconductors. For example. SiC MOSFETs principally provide fast switching speed with low switching losses. The commercial SiC modules use simple and reliable wire bonding technology with wider energy bandgap. Such physical characteristics equip the devices with an improved overall efficiency by providing high critical breakdown field and thermal conductivity than their conventional counterparts [4] [5]. Due to the exceptional performance made possible by silicon carbide power devices, control electronics are required to operate in the same demanding conditions. The reduced switching power losses allow high frequency operation ultimately and reduce the overall system cost [6]. The fast switching is compulsory for avoiding long transients of the MOSFET ohmic region operation. In the on time, a positive gate drive voltage should be applied to keep the on-resistance as low as possible.

2.1　SiC MOSFETs structure

The SiC MOSFET works as a switch in such a way that the gate controls the current flowing from drain to source. In a typical SiC MOSFET structure, two opposite p-n junctions make source-to-body junction and drain-to-body junction. The junction blocks the voltage either from drain to source or source to drain as shown in Figure 2. The BJTs bipolar devices enhance the conduction between collector and emitter by injecting the minority carriers through their base terminals. While in the MOSFETs, gate terminal is gate-oxide insulated. However, the negative charges start accumulating on the gate terminal when a positive voltage with a reference to the source is applied to the gate terminal. Ultimately, a conduction channel forms between drain to source allowing the drain current to pass between the two terminals [7] [8]. SiC MOSFETs usually come in TO-247 and TO-247-4 packages. TO-247 has three terminals such as Gate (G), drain (D) and source (S). While TO-247-4 comes with an extra terminal to have a shorter path to the gate called kelvin source.

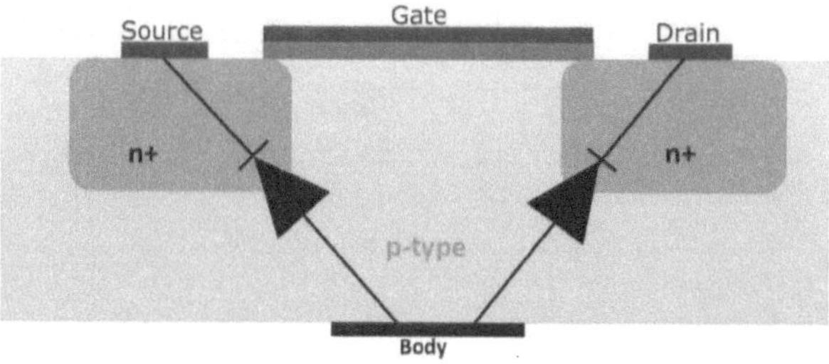

Figure 2 Internal structure of MOSFET

2.1.1 Intrinsic resistance and capacitance in SiC structure

The detailed n-channel SiC MOSFET structure showing its intrinsic resistances is shown in the Figure 3.

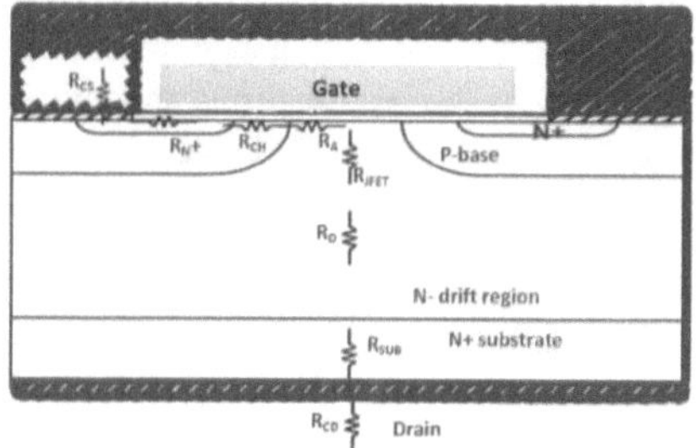

Figure 3 Intrinsic resistances present in the N channel MOSFET

The drain to source resistance $R_{DS}(on)$ of the SiC device is the sum of present intrinsic resistances in the model as shown in the equation (1.1).

$$R_{DS(ON)} = R_{CS} + R_N + R_{CH} + R_A + R_{FET} + R_D + R_{SUB} + R_{CD} \qquad (1.1)$$

The intrinsic resistances are temperature-dependent, and capacitances are temperature-independent. This means that the conduction losses depend on the temperature while switching losses of the MOSFETs are rather independent of temperature as switching speed relates to charging and discharging of the input capacitance (C_{iss}). However, the gate-to-drain capacitance (C_{GD}) and the gate-to-source capacitance (C_{GS}) depend on are voltage applied [31]. Input capacitance (C_{iss}) is the sum of C_{GS} and C_{GD} and directly influence the switching speed [10].

The equivalent circuit for the intrinsic capacitances in an n-channel MOSFET is shown in Figure 4.

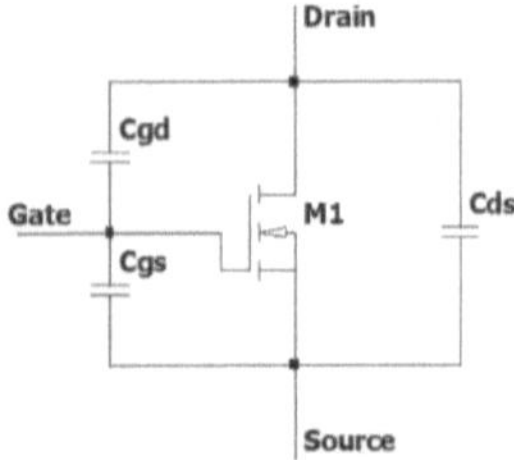

Figure 4 Equivalent Intrinsic capacitance present in the N-channel SiC MOSFET

Output Capacitance (C_{oss}) is equivalent to the sum of drain-to-source capacitance (C_{DS}) and gate-to-drain capacitance (C_{GD}) and influences the resonance of the circuit. Reverse transfer capacitance (C_{rss}) is the gate to drain capacitance (C_{GD}) and often referred as Miller capacitance. C_{rss} affects the voltage rise and fall time during switching [11][10]. The Figure 5 below shows the nonlinear curves for input and output capacitance vs drain to source voltage in SiC MOSFET. It can be verified that the input capacitances C_{GS} and C_{GD} are heavily drain to source dependent [11].

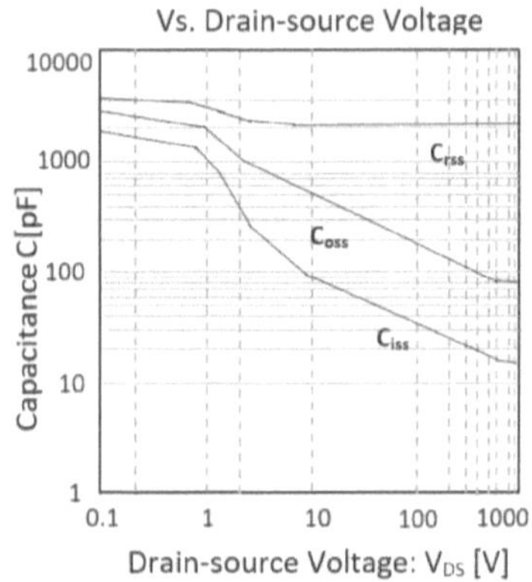

Figure 5 SiC MOSFET Capacitance vs V$_{DS}$ curve

C_{iss} must be charged in order to drive the MOSFETs. Gate-drain capacitance (C_{GD}) also called as miller capacitance is nonlinear function of voltage and is equivalent to C_{rss}. It is one of the most important parameters providing a feedback loop between input and output of the circuit. C_{iss} is equivalent to the sum of the C_{GD} and C_{GS}. If C_{oss} is large, a current arising due to C_{oss} flows through the output even under the turn-off condition. Q_{GD} is the charge amount necessary to drive (charge) C_{RSS}. Altogether, all these parameters greatly affect switching speed [7].

2.2 Silicon vs SiC MOSFETs

Si MOSFETs and IGBTs have been used in power converters for a long time. However, SiC MOSFETs have emerged as a new technology, showing benefits surpassing those devices given their intrinsic material properties. As comparison to Si based devices the SiC offers superior advantages like wider energy bandgaps, higher critical electric fields with thinner blocking layers, higher electron saturation velocity, and higher thermal conductivity. The higher bandgaps result in much lower current leakages and higher temperatures [13]. While higher critical electric fields and higher doping concentrations helps to come up with a thinner blocking layer. Higher electron saturation velocity gives rise to higher frequencies. While the improved thermal conductivity helps in heat dissipation, thus the devices can work on higher power densities. These characteristics are summarized in Figure 6. The material properties of SiC directly translate into system-level advantages over systems using Si devices, including reduced size, cost, and weight. Consequently, SiC MOSFETs are increasingly replacing Si power devices [12].

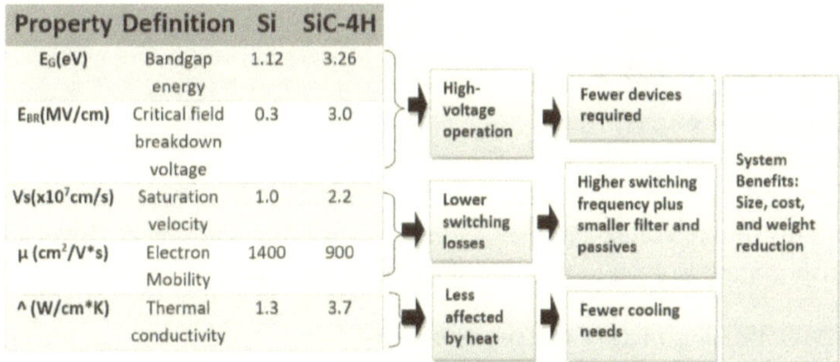

Property	Definition	Si	SiC-4H
E_G(eV)	Bandgap energy	1.12	3.26
E_{BR}(MV/cm)	Critical field breakdown voltage	0.3	3.0
Vs($\times 10^7$ cm/s)	Saturation velocity	1.0	2.2
μ (cm²/V*s)	Electron Mobility	1400	900
^ (W/cm*K)	Thermal conductivity	1.3	3.7

Figure 6 Major comparison between Si and SiC

2.3 Important parameters of SiC MOSFETs

Although SiC MOSFETs exhibit the similar general switching behaviour like traditional silicon MOSFETs, there are many considerable design requirements associated with the device characteristics. Recently, many industrial SiC MOSFETs offer great features including low $R_{DS}(on)$, very low switching losses, threshold-free on state characteristics, fully controllable dV/dt, and commutation robust body diode. SiC MOSFETs are known for their lower losses by providing the faster rise and fall time. Previous studies showed that these fast slew rates lead to problems such as electromagnetic interference (EMI) [15] [13]. Some SiC MOSFETS also come with kelvin source pin which effectively reduces the parasitic inductance of the source lead of the power MOSFETs. Some important characteristics of widely available SiC MOSFETS are shown in the Table 1 taken from the data sheets of manufacturer.

Table 1 SiC MOSFET overview

Manufacturer	Part-ID	V_{rated} (KV)	$R_{DS(on)}$ (mΩ)	I_D (A)	V_{GS} (V)	$V_{GS(th)}$ (V)	Ciss (pF)	Coss (pF)	Q_g (nC)	R_g (Ω)
ROHM	SCT2080KE	1.2	80	40	-4V,+22V	2.8	2080	77	106	6.3
CREE	CMF20120	1.2	80	42	5V,25V	3.2	1915	120	90.8	5
Little-Fu	LSIC1120E0080	1.2	120	25	-6V, +22V	2.8	1700	82	92	0.85
Infineon	IMZ120R045M1	1.2	45	52	-10V, +20V	3.5	1900	115	52	4
CREE	C3M0075120K	1.2	75	30	-4V, +19V	2	1350	58	51	10.5

2.4 IV-Characteristics

The gate-to-source voltage (V_{GS}) greatly influences the drain-to-source resistance $R_{DS}(on)$ of the MOSFET, and eventually the drain current (Id) as seen in the Figure 7. The field effect of the gate is increased due to high gate-to-source voltage applied. Hence, the on-state resistance is inversely proportional to gate-to-source voltage $V_{GS(ON)}$.

A MOSFET lies in its ohmic region when V_{DS} influences the drain current, but only when drain-to-source voltages is low. In the active region of the MOSFET, I_D is independent of drain-to source voltage and only depends on the gate-to-source voltage. Cutoff region of the MOSFET is the last region where all the drain current is blocked [8] [10].

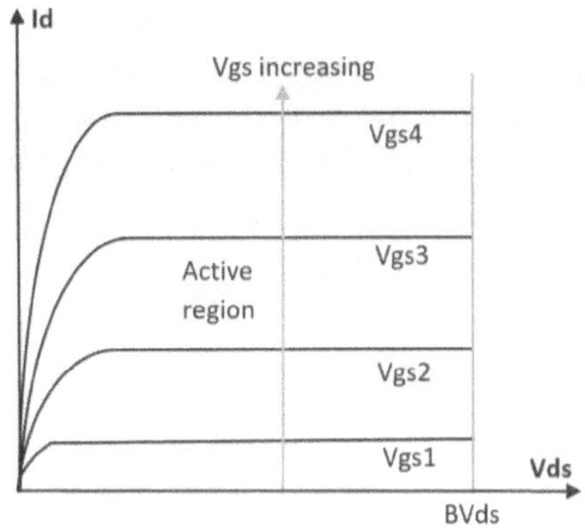

Figure 7 SiC MOSFET I-V characteristics

As the field effect from the gate terminal is not high enough to induce a conducting channel between drain and source because of low gate to source voltage, thus blocking the Id.

2.5   SiC MOSFET Switching Characteristics

The typical switching characteristics of SiC MOSFETs are similar to their traditional Si counterparts. A typical circuit application for SiC MOSFET is shown in the following Figure 8. The load is purely inductive thus represented by I_o.

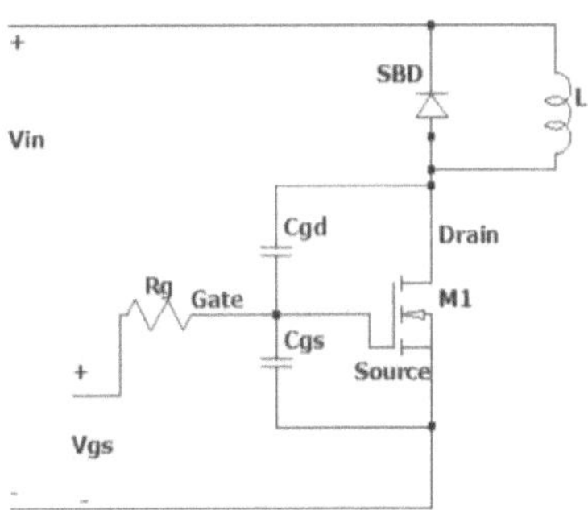

Figure 8 MOSFET example in typical inductive load switching

10

2.5.1 Definition of switching characteristics parameters

The typical switching transients for the hard switching applications with important parameters are shown in the Figure 9. It shows the turn-off and turn-on transients with important switching times and derivatives parameters.

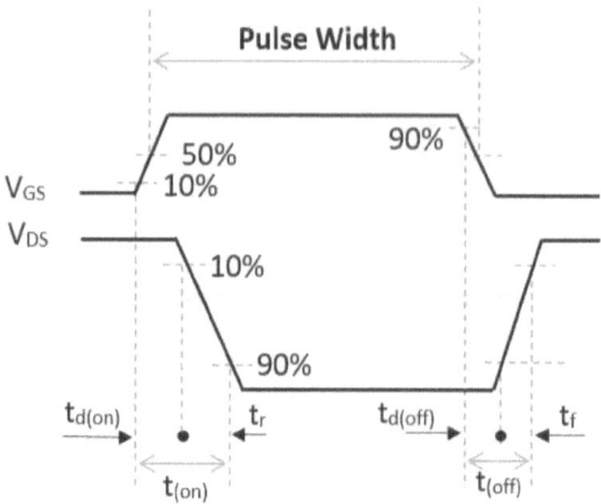

Figure 9 SiC MSOFET switching parameters

Where $t_{d(on)}$ and $t_{d(off)}$ represent the delay, when the gate to source voltage V_{GS} reaches to 10% and 90% of its final value with respect to the drain to source voltage. The tr and tf are the rise time and fall times for drain to source voltage, respectively. The $t_{(on)}$ exhibits the 10% of V_{GS} to 90 % final value of V_{DS}.

2.5.2 Turn-On Switching Characteristics

The switching transients of the MOSFETs include turn-on event together in a hard switching typical application as shown in the figure below. Firstly, it is assumed that the MOSFET is in its off state and Io is freewheeling in the Schottky barrier diode (SBD). The load is assumed to be purely inductive and MOSFET is supplied with a positive gate to source voltage, as a result the turn-on transient will initiate.

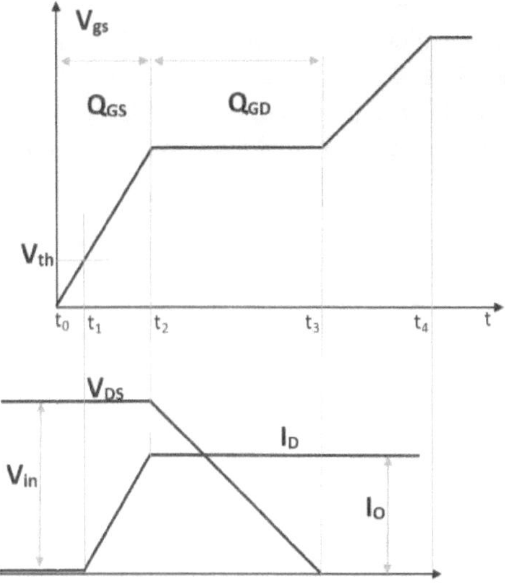

Figure 10 MOSFET turn-on transient

With the rise in gate to source voltage, the capacitance C_{GS} will be charged by the gate current through gate resistance Rg. By charging the gate capacitance C_{GS} of the MOSFET, the field effect from the gate induces a conducting path from drain to source and drain current starts to rise. The $Rds_{(on)}$ will decrease as the V_{GS} increases until it reaches the load current (t2). At point t2, capacitance C_{GS} is fully charged and V_{GS} levels a point called Miller plateau [7].

During this period, the gate to drain capacitance C_{GD} is charged by gate current I_G and the drain to source voltage starts falling towards zero. At the interval t3, the drain to source voltage only depends upon the on- state resistance Rds(on) and the period is known as ohmic region [10]. The gate voltage of the MOSFET continues to rise until reaching the final gate voltage. In the real application, MOSFET turn-on transient includes the reverse recovery of the diode present within the MOSFET. The reverse recovery event occurs at the time t2 due to the reverse recovery current (I_{rr}) present at the diode turn-off transition. Increase in the drain current due to I_{rr} leads to a slight bump in the gate voltage (t_{rr}). Thus, the turn-on switching losses includes the diode reverse recovery losses [11][10].

2.5.3 Turn-Off Switching Characteristics

In the turn-on transition, the freewheeling diode is reversed biased when the load current I_0 is flowing through the MOSFET. Now if the gate voltage (V_{GS}) is pulled down to zero or negative level, the turn-off transient will begin. The turning off event will involve exact same transition in reverse order. The ideal switching behavior for turn-off is shown in the Figure 11.

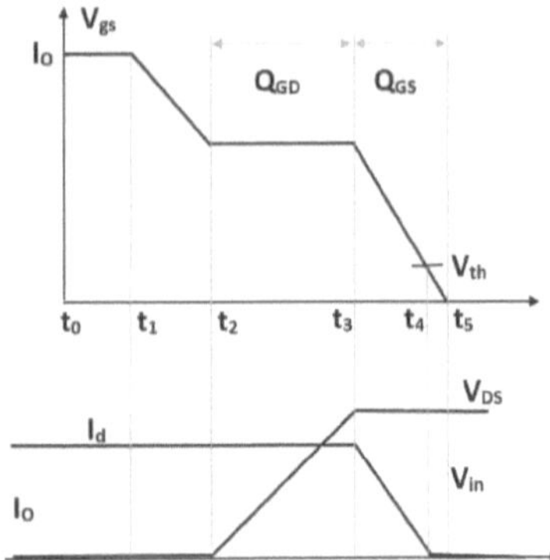

Figure 11 MOSFET turn-off transient

The actual turn-off switching events often show considerable overshoots in the voltage and currents at the time when V_{GS} crosses the miller plateau (t3) as shown in chapter 5.4. This effect happens due to stray inductance Ls resonance in the circuit causing the high dI/dt and overshoot in drain to source V_{DS}. Such overshoots in current and voltage significantly influence the turn-off switching losses [3].

2.6 Influence of gate driver on SiC MOSFETs

Gate driver works as a power amplifier which can convert the low input power from the microcontroller to high current drive output. The SiC MOSFETs offer a lower transconductance than Si MOSFETs with almost same V_{GS}, and thus the channel resistance decreases significantly with V_{GS}. It is recommended to maximize the V_{GS} to a suitable value for minimizing the conduction losses. However, as seen in the previous studies, SiC transistors demand a rapid transition in gate voltage without significant ringing and overshoots [13]. In comparison to conventional IGBT's and MOSFETs, the overall quality of the gate to source voltage waveforms has a large impact upon the device performance and, therefore, worthy of attention.

To achieve a faster transition for the gate voltage with a required gate current these advanced SiC devices need a better and robust gate driver design. The overall gate driver function is to charge and discharge the gate capacitances C_{iss} and C_{oss} by providing a required sink and source current to the MOSFET rapidly. To drive advanced SiC devices on target requirements, the gate driver must perform turn-on and turn-off transition safely.

2.6.1 Turn- On

The path of the source current of SiC MOSFET is shown in the Figure 12. The turn-on transition needs a large peak to peak current, efficient to charge the SiC internal gate capacitance and fast enough for minimizing the switching loss [17] [15]. The entire turn–on event occurs within approximately few ns for a full V_{GS} swing when ΔV_{GS} = 15 V and calculated C_{iss} = C_{GS} + C_{GD} = 1000 pF, yielding a required peak current I_G =1.5 A, according to the following equation:

$$I_G = \frac{(C_{GS} + C_{GD}) \cdot \Delta V_{GS}}{\Delta t} \tag{1.2}$$

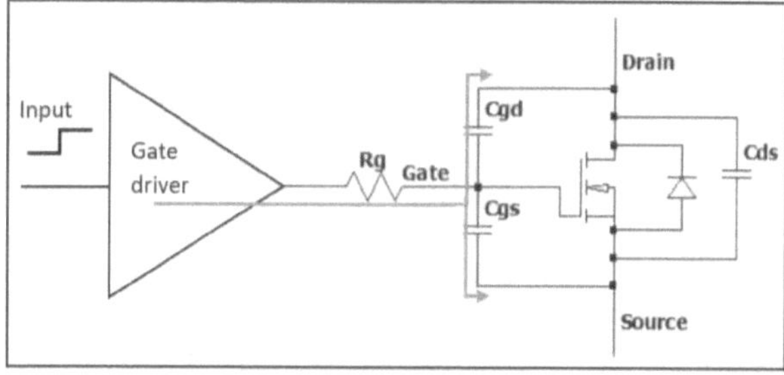

Figure 12 SiC MOSFET source current

2.6.2 Turn-Off

Turn-off transition occurs in reverse order exactly of turn-on state as mentioned in previous section 2.6.1. Gate drive circuit sinks a large amount of the peak current which ultimately discharges the C_{GD} and C_{GS} capacitance of the SiC MOSFET as fast as possible. Moreover, the gate driver impedance during turn-off should be minimum to hold the MOSFET gate low [16]. It can be problematic due to the low threshold voltage (V_{th}) associated with the SiC MOSFETs. It is compulsory that the SiC gate being pulled below ground and the capability of the sink current should be higher than the source current capability [19]. The flow of IG_{sink} is highlighted in the Figure 13.

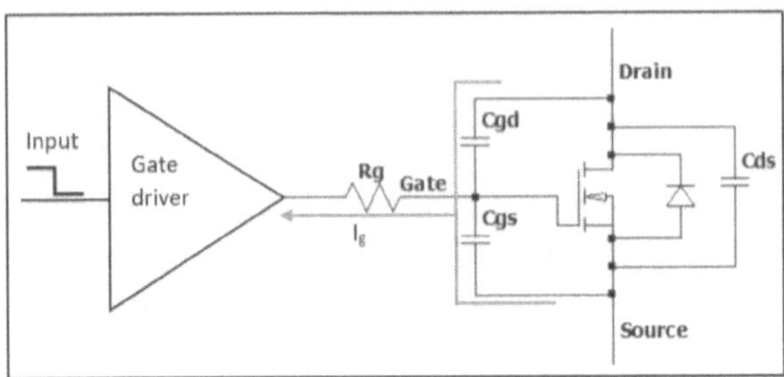

Figure 13 SiC MOSFET sink current

2.6.3   Driver strength (I_G)

The gate current or drive strength controls how fast the device's input capacitors charge and discharge. When the gate current increases, the switching losses decreases and vice versa. The required gate-drive strength depends on gate charge of the SiC MOSFET (Qg) as shown in Figure 14, gate current is calculated as in a time of ton is calculated as:

$$I_G = \frac{Q_G}{t_{on}} \tag{1.3}$$

This current is the average current required to turn the device fully on. The region of interest, however, is the Miller plateau region, where the gate voltage is constant during the switching transient [16].

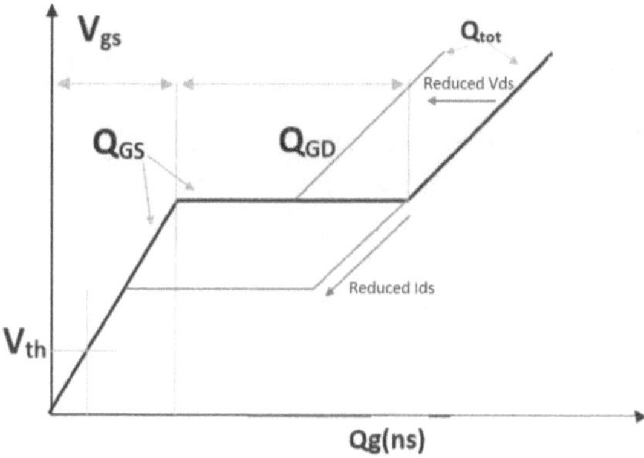

Figure 14 Gate capacitances of SiC devices

2.6.4 Propagation delay

Another important part in gate driver technique is to minimize the losses due to dead time and propagation delay. Higher frequency also means higher switching losses, having a higher propagation delay or dead time means that current circulating through the diodes in the loops and resulting in more losses overall. Propagation delay is one of the key parameters of a gate driver that can affect the losses and safety of high-frequency systems. Propagation delay is defined as the time delay from 50% of the input to 50% of the output, as shown in Figure 15.

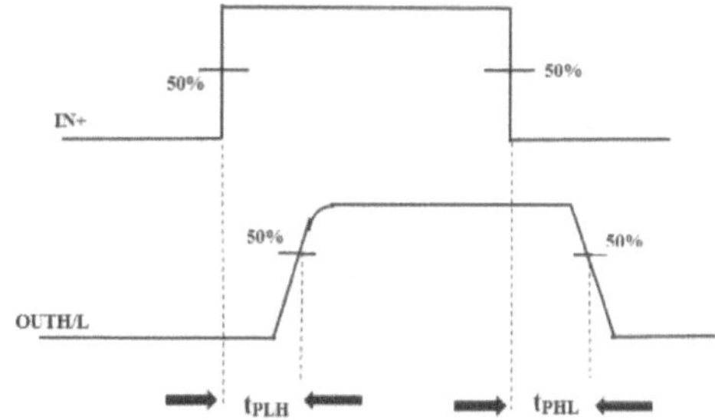

Figure 15 Pulse with distortion

This delay affects the timing of the switching between devices, which is critical in high-frequency applications where the dead time is necessary so that two devices do not turn on at the same time, which would cause shoot-through and reduced efficiency [7]. If the dead time is smaller than the propagation delay, then both devices will turn on at the same time.

2.6.5 Threshold Voltage and Source Inductance

One of the most critical challenge faced by the designers of SiC MOSFETs is to control the gate threshold voltage. Early SiC MOSFETs had problems with threshold stability. The threshold shifts due to soaking at a bias condition. The continuous operation with a negative and positive gate bias at an elevated temperature usually results into a threshold shift [4]. These problems, however, have been resolved in the latest generations of SiC MOSFETs [17]. The threshold voltage is a compromise by design between noise immunity and $R_{DS(on)}$. The temperature dependence is nonlinear, so the specified threshold is generally close to 2 V at 175 °C [11]. These properties make the SiC transistors more sensitive to unintentional turn-on. To overcome this issue, a negative voltage at gate is often necessary to confirm a safer turn-off transition. The SiC transistor usually does not show a flatter Miller plateau but manifest a fine transition between the regions of the ohmic to saturation transition. Thus, it is recommended to have a higher voltage for gate around 18-20V [21] [3] [18].

It is highly suitable to use a negative gate voltage with the SiC MOSFETs in the high frequency switch mode power supplies. The gate to source voltage is designed below ground level to keep the device OFF when there is no active switching. The major reasons for using negative voltage are source inductance and gate drive impedance. Source inductance is the inductance introduced by the output current loop and gate driver current loop [15]. The gate drive voltage combined with the source inductance present in the loop have a significant effect on the higher frequency applications under the load. This is due to the source lead inductance which couples the output switching current back to the gate drive and ultimately slow the gate drive. Gate drive impedance is important so that the gate driver provides a lower impedance path in off-state other than the on-state drive on the opposing transistor. This can limit the Miller effect and cross conduction losses in the bridge configuration [13]. Thus, a gate driver with a negative voltage option plays an important role in reducing the cross-conduction losses. The detail study and effects of the negative voltage will be presented in chapter 7.2.1.

2.6.6 Miller Effect and Gate Drive Impedance

Due to the high-power levels and fast-switching devices, high dV/dt and dI/dt occurs at each switching instance. The circuit and switching devices contain parasitic capacitances and inductances which further interact with transients causing destruction of the system [18] [25].

In bridge configurations, during the fall times of V_{DS} of one switch, the current is pushed towards the gate by Miller capacitance C_{GD} of the complementary switch. A spike in V_{GS} commonly appears due to the high dV/dt by the Miller current. In case when V_{GS} exceeds the threshold voltage due to the rapid negative slew rate, a shoot through in the bridge can occur as shown in the Figure 16. This is more common in the low threshold voltage power transistors specially SiC MOSFETs [18] [23] [24].

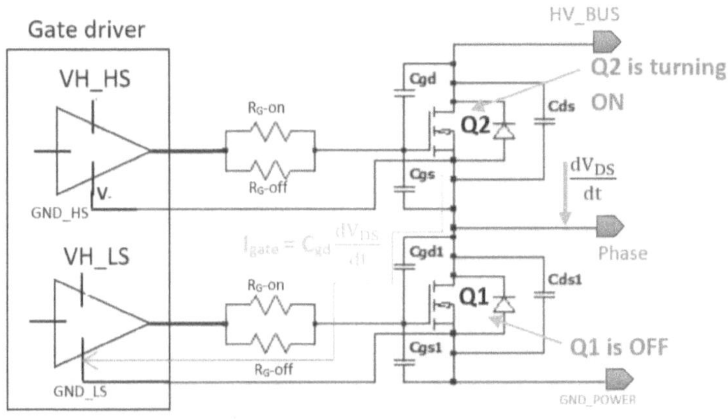

Figure 16 Miller effect seen in bridge configurations

In general, duality glitch phenomenon usually occurs due to the negative or positive dV/dt on the switching nodes resulting in the flow of I_{GD} through the capacitance C_{GD}. The rapid change dV/dt induces higher trumped-up gate voltage causing a shoot through in phase leg. Additionally, some of the current flowing through the C_{GD} flows out through the gate terminal and the gate driver sink resistance produces a spike in the gate voltage.

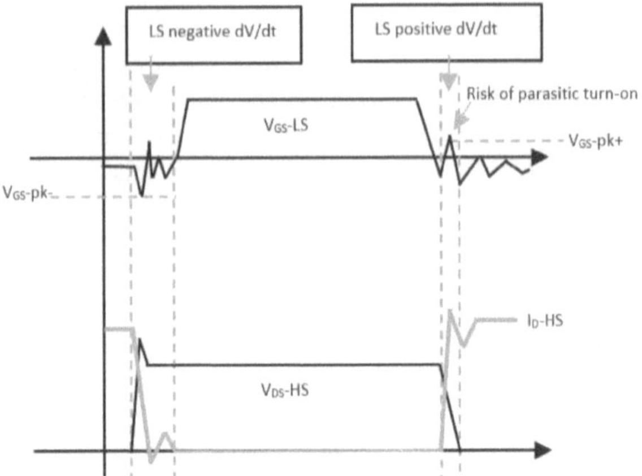

Figure 17 Parasitic turn-on in bridge configurations

This can drive the upper device back into conduction and can cause a short through. A typical illustration causing the spike is depicted in the Figure 17. There are different solutions to overcome this issue during the transistor turn-off transition. In general, limitation of gate loop and common source inductances can minimize this false turn on phenomenon as explained in section 7.2.1. A small capacitor between the gate and source can be another option but will cause high losses overall and significant efficiency decrement. Moreover, selecting a gate driver with negative gate voltage option and low pull-down impedance path proves a significant improvement in the Miller avoidance [20]. One of the solutions is to choose a correct gate resistor ratio which can be 2:1 or 4:1. The reduction of Rg-off solves the issue to some extent. However, a smaller Rg-off might lead to extremely faster switching and very high stress on the transistor by generating the higher dV/dt. The higher dV/dt effect can be also limited by choosing a higher gate resistor for the transition.

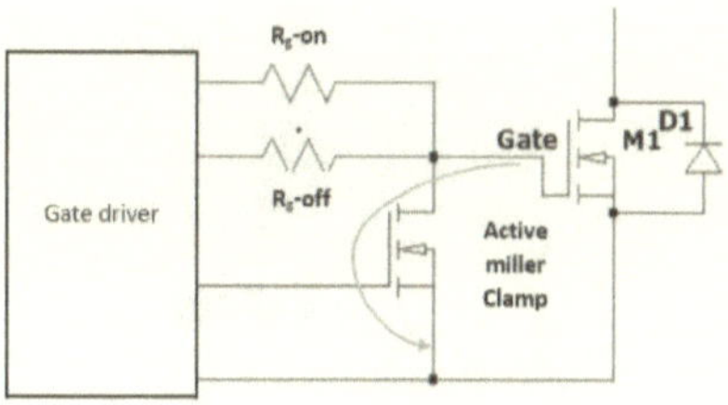

Figure 18 Active Miller clamp function in gate driver IC's

The other solution is more practical by adding the external pull down MOSFET to the gate which can bypass the Miller current during the turn-off. This solution is called active Miller clamp. The Miller clamp contains an extra MOSFET within the gate driver circuit as shown in Figure 18. This clamp circuit is controlled by the gate driver IC. The bypass MOSFET is normally ON when the gate to source voltage of M1 is reduced to the threshold voltage. This method is more practical as this will not affect either the switching time or dV/dt.

2.7   Clamped inductive load in MOSFET switching

In power electronics different topologies can be derived from a circuit consisting of two switches connected in series or parallel. The switches Q1 and Q2 can be used or replaced with internal body didoes D1 and D2 depending on the application requirements.

This structure is implemented in many converter applications such as boost and half-bridges and is called as H-bridge circuit. Multiple copies of the H-bridges can be used to form a 3-phase inverter or full-bridge dc-dc converter. Here, blocking is required inside orientation, and diode conduction in the other direction is necessary for the proper operation. This operation can be obtained by activating or deactivating the switches, and this method is called as 'synchronous switching'. Switch mode power supplies cast an inductive effect to the switching node (Vsw) between the two switches. This node can be further connected to an inductor forming a clamped inductive load. This specific set of the system has unique system dynamics. During the Q2, inductive load charges the inductor from the one side. While in the Q1 turn-on state, the inductor will be charged in the opposite manner.

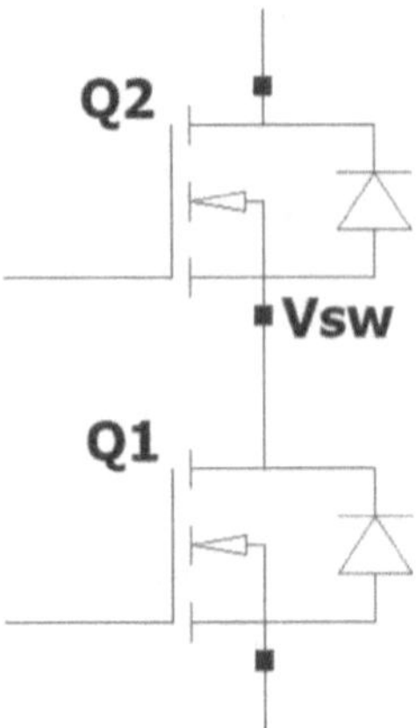

Figure 19 H-bridge configuration in power electronics

In an ideal case where the load is purely inductive, the DC voltage V_{bus} is applied to the load L, Q1 turns on and the inductor current will start to increase as given in equation 1.3. Q1 turns off and I_L freewheels though the diode V_{D1}. Here, the I_L starts to reduce slowly.

$$I_L = V_{DC}\,\frac{t_1}{L} \tag{1.4}$$

At the turn-off transition, the inductor adds the voltage of the output point node though the high side freewheeling diode. The current at this point is the full inductor current, and only starts to decrease when the diode is in the forward-biased condition. The voltage and current across the MOSFET terminals rise to V_{BUS} and I_L. At turn-on transition, this process is reversed, the current from 0 to I_L, while at voltage at V_{BUS}, the current reaches to I_L and the voltage decreases. This is called 'hard switching' and usually involves the high peak power. There are other techniques to reduce the voltage or current stress across the switch which are called zero current switching (ZCS) and zero voltage switching (ZVS) [22].

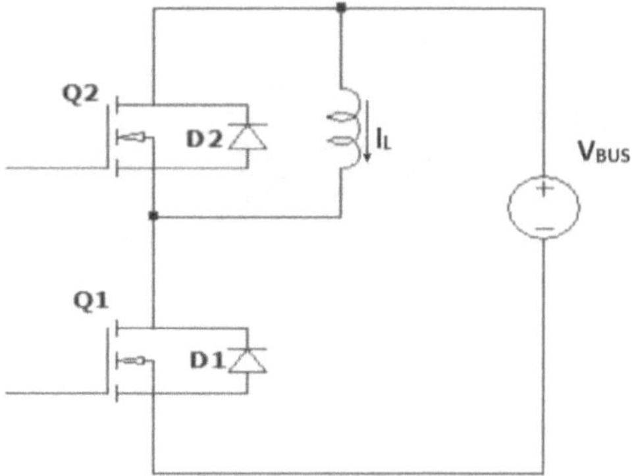

Figure 20 Clamped inductive load example circuit, with optional synchronous MOSFET Q2

The clamped inductive load can be further explored by replacing the theoretical switches with the real components as mentioned in the section 4.4. For such experiment, a MOSFET can be considered with an inductive load in the hard switching application. The idealized waveforms with clamped inductive load are shown in Figure 21.

The hard switching event with a clamped inductive load can be described in various phases, it begins with the V_{GS}, changing from zero or negative state to V_{th}. After V_{GS} reaching the threshold limit, the MOSFET starts to conduct current. The current starts to follow through the drain to source and drain to source voltage at this point remains constant. If the MOSFET drain current I_D is less than the inductor total current, remaining inductor current will have an additional path through the diode which is parallel to the load inductor. The drain voltage at this point is the voltage drop across the diode and remains above the V_{bus}. The gate to source voltage will continue increasing until the drain current is equal to the inductor current [21]. As the drain current crosses inductor current, this makes the diode to come in reverse biased condition and drain voltage starts falling and MOSFET is in completely on-state.

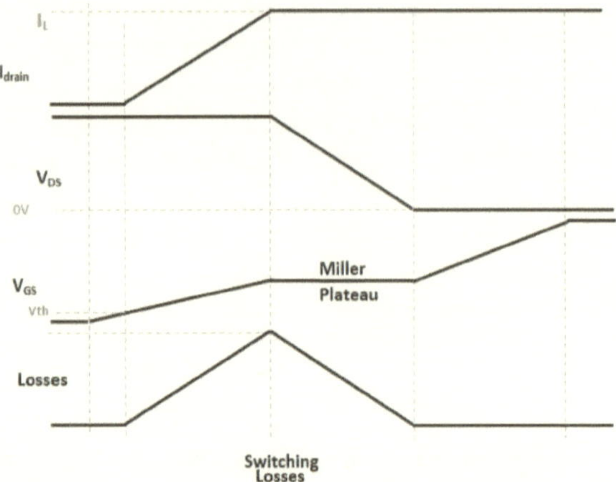

Figure 21 Ideal MOSFET switching waveforms for a clamped-inductive load turn-on

It is important to note that the clamped-inductive switching configuration develops several peak conditions during switching that do not exist to the same extent in other common configurations. The clamped-inductive load guarantees that the MOSFET will experience maximum di/dt, dV/dt, and instantaneous power in a single controlled switching event [8]. While several performance maximums will occur, the system will constrain peak MOSFET drain currents, and the total energy dissipated is a function of the switching speed.

2.8 Switching losses in MOSFET

Conduction losses and the switching losses are amongst the major losses in half bridge configuration in MOSFET. The conduction losses occur due to on-state resistance of the power MOSFET. The drain to source voltage across the MOSFET is not zero due to the static conduction losses.

$$P_{cond} = Id_{rms}^{2} \times Rds^{ON} \tag{1.5}$$

Where the I_D is drain current through the switch. The $R_{DS\text{-}ON}$ is the on-state parasitic resistance of the MOSFET. The ideal switching waveform is shown in the Figure 22. At lower frequencies, static conduction losses are more dominating and can be averaged over whole period.

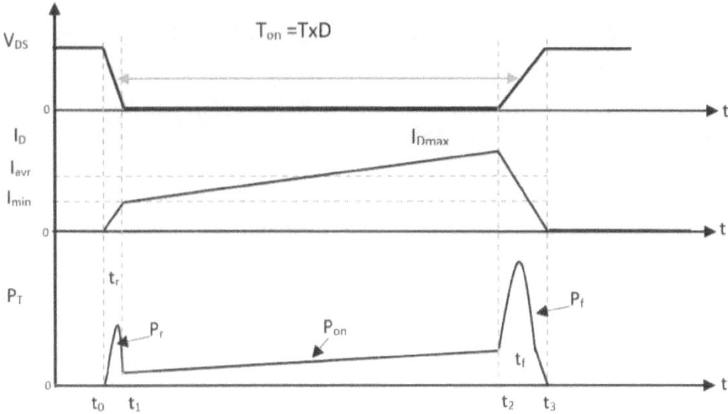

Figure 22 Power transistor losses due to switching times and R_DS(on)

$$Id_{rms} = \sqrt{D}\left(\frac{Id_{min}+Id_{max}}{2}\right) \qquad (1.6)$$

Where Id_rms is the RMS current through the switch and D is the duty cycle. The switching losses can be divided in the rise and fall times for the MOSFET as mentioned in equation 1.6. The switch losses during rise time can be defined as:

$$P_r = \frac{1}{T}\int_0^{tr} V_{DS}\ I_d\ dt = \frac{1}{T}\int_0^{tr} V_{IN}\ I_{dmin}\left(1 - \frac{t}{t_r}\right)dt = \frac{V_{IN}.I_{dmin}\ t_r}{6T} \qquad (1.7)$$

Where V_in is the input voltage and tr and tf are rise and fall times, respectively.

The switch losses during on transition:

$$P_f = \frac{1}{T}\int_0^{tf} V_{DS}\ I_d\ dt = \frac{1}{T}\int_0^{tf} V_{IN}\ I_{Dmax}\left(1 - \frac{t}{t_r}\right)dt = \frac{V_{IN}.I_{D\ max}\cdot t_f}{6T} \qquad (1.8)$$

The gate charge losses can be defined as:

$$P_G = (Q_{g-H} + Q_{g-L}).V_{gs}.f_{sw} \qquad (1.9)$$

$$P_G = (C_{g-H} + C_{g-L}).V_{gs}{}^2.f_{sw} \qquad (1.10)$$

Where the $Q_{g\text{-}H}$ and $C_{g\text{-}H}$ are the MOSFET gate electric charge, charge capacity and f_{sw} is the switching frequency.

Dead time losses can be defined as:

$$P_D = V_{DD} \cdot I_D \cdot (t_{Dr} + t_{Df}) \cdot f_{sw} \tag{1.11}$$

Where the t_{Dr}, t_{Df} are the rising and falling dead times and V_{DD} is the diode forward voltage of the lower MOSFET.

The turn-on and off energies can be calculated as:

$$E_{on} = \int_0^{ton} Vds.Id(t)dt \tag{1.12}$$

$$E_{off} = \int_0^{toff} Vds.Id(t)dt \tag{1.13}$$

$$P_{SW} = \left(E_{on} + E_{off}\right) \times f_{sw} \tag{1.14}$$

The overall switch losses are the sum of losses during all and can be specified as

$$P_{tot} = P_{cond} + P_r + P_f + P_G + P_D \tag{1.15}$$

2.9   Gate Driver for SiC MOSFETs

A gate driver is the last and crucial stage of circuitry where the control logic determines different state of the MOSFET through the gate of the MOSFET. A simple microcontroller is only able to deliver somewhere 5V and few milliamps where these SiC devices need higher current to switch 15-20V at gate terminal to minimize the switching losses. Where traditional drivers give low output drive and big latency in the switching patterns. As performing at elevated voltages and current levels require additional considerations like protection and isolation. The answer is to go with faster and more reliable design approach in the drivers.

A simple gate drive circuit can be built out of discrete components to drive a power MOSFET. Two simple BJTs or MOSFETs forming a totem pole or push-pull configuration can increase the drive strength as shown in Figure 23(a). But these circuits distributed on PCB with longer PCB traces will cause extra stray inductances in the gate loop [24]. A simple and more suitable approach can be gate driver IC, which combine all these circuitries on a micro scale thus excluding the extra inductance problems in the gate drive section. These gate drivers come with extra features incorporated within IC chips as level shifting, short circuit protection and many more.

More advanced gate driver type is the current source gate drivers which offers another solution for driving the power MOSFETs. The current source gate drivers use a charged inductor as a primary current source to charge and discharge the gate capacitances of the MOSFET as shown in the Figure 23(b). The current mode drivers offer two basic modes for operation which depend on the current stored in the drive inductor and the required energy to drive the MOSFET. The inductor requires the charging before turn-on and full discharging before the turn-off transition which increase the propagation delay from input to output. This solution is presented in previous studies where it provides good current drive for the Miller plateau using the same energy stored in gate capacitances but results in increased control latency and needs a more complex design approach [25].

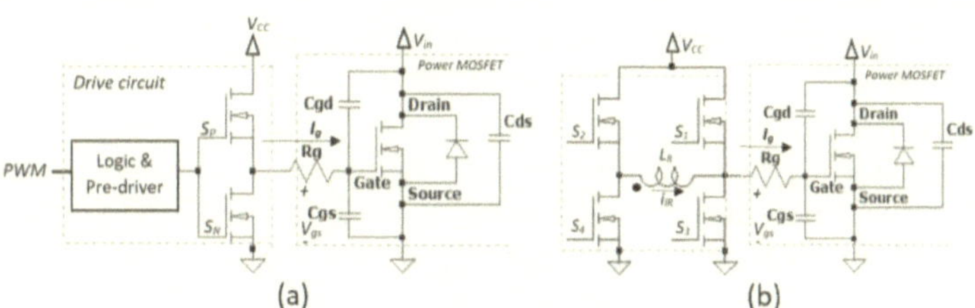

Figure 23 (a) Voltage source gate driver (b) Current source gate driver

The gate driver connection to the MOSFET is another important design parameter [26]. The simplest method is a direct connection from gate driver output to the gate terminal of MOSFET. The latest gate driver ICs in the market can provide the high gate current thus require a modification before the gate of the power MOSFET as shown in the Figure 24(a). A gate resistance Rg can be added next to the gate driver IC output to reduce the output current. By removing the current limiting resistance, gate current is only limited by the internal gate resistance of the power

MOSFET or gate driver. The Rg adjustment requires an additional verification that the gate driver should provide sufficient but not excessive drive. The excessive drive can result in maximum dV/dt and overshoots in the drain to source voltage [20]. These overshoots can cause excessive stress to the power MOSFET. Fundamentally, the overall performance of the system is heavily gate driver dependent.

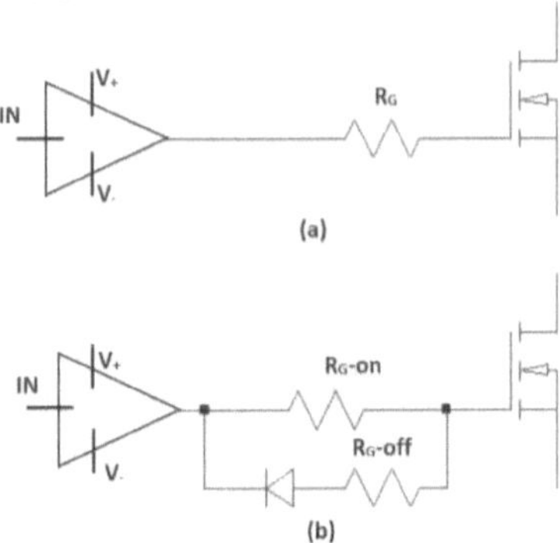

Figure 24 The gate driver current limiting solutions (a) turn-on (b) turn-off

Some gate driver IC's in the market only provide the single output channel instead of split outputs for turn-on and turn-off. For this approach, a diode in parallel with the separate pull-down resistor can be implemented in the design. This method can provide the different drive strength for turn-on and turn-off transition as shown in the Figure 24(b). The additional path provided by diode is used to short circuit the resistance for turn-off transition creating a lower impedance path for the current. Such solutions will result in higher noise immunity but will cast more complexity to the design and increase the overall cost [18].

3 Methods

The manufacturers usually provide the datasheets for the SiC MOSFETs. As these devices are new and often show variability in the characteristics thus, we firstly preferred to characterize these devices to know their actual behavior. This is very important especially in the development stage. The project was carried out according to the following steps.

3.1 Double pulse test

To match the concrete and verifiable goals for the project a double pulse test (DPT) setup is designed. The DPT tester demonstrates the gate driver driving the SiC MOSFETs and effects related to switching performances. The setup is designed to test different SiC MOSFETs, SiC Schottky barrier diodes, and different gate driver circuits. Wide band gap devices especially SiC MOSFETs need prudent testing design. For example, circuit layout with minimum parasitics in the design is required for the accurate transients measurements. Further design layout is mentioned in the chapter 4.4.

3.2 Evaluation of Gate drivers

The SMD components on larger printed circuit board result in big current loops in the design influencing the driving performance of the gate driver. The commercial gate driver ICs come in small PG-DSO package with several performance benefits. Thus, it is always best to choose a commercial gate driver IC. Thus, four different gate drivers with superior performances are selected and optimized with decoupling capacitors and suitable power supply. The focus here is to design a gate driver in close proximity to the SiC MOSFET to minimize the inductance produced by long current paths [18]. The various gate drivers, their specifications and design are mentioned in 4.1.

3.3 Evaluation of SiC MOSFETs with gate drivers

One of the goals of the project is to investigate the SiC MOSFETs performances and integrate these devices into the required application. SiC MOSFETs in the market need to be investigated by datasheets and the result obtained from the DPT. The optimal target is to choose a device which offers balance between fast and clean waveforms. SiC MOSFET needs to be compared with the same gate driver to observe the performances of both devices as mentioned in the section 5.2.

3.4 Simulation Models

A double pulse test simulation circuit is modeled in the LTSPICE to check the reliability of the gate drivers and SiC MOSFETs and results are compared between DPT and simulation. The common parasitics such as PCB parasitics and common source inductances are added in the simulation software. Some SiC MOSFETs manufacturers provide the spice models for their devices, thus these models are further investigated in the simulation circuit as described in the section 5.11.

3.5 Evaluation of drivers in DC-DC converter

After obtaining the evaluation of the drive circuits in DPT, a DC-DC converter in phase leg configuration is built. The performances of drive circuit can be different in the DPT and converter due to rapid change in currents and voltages in the converter. As the SiC MOSFETs usually operate at very high voltage and current levels, the driver actual performance is monitored in the converter as described in section 7.3.

4 Implementation

4.1 Gate driver ICs

The main characteristics of driver ICs include the rise t_r and fall t_f times, propagation delay time, peak output current Ipeak, and desaturation threshold V_{Desat}(th). The considerable features of gate drivers are active Miller clamping, overcurrent protection, under voltage lockout (UVLO), and soft turn-off. Overcurrent protection should cause a quick response as it uses the desaturation detection when the device goes through short circuit and consequently the driver will force it to the desaturation mode. Finally, an active miller clamp circuit should be added for preventing false turn-on of the SiC MOSFET. Moreover, gate drivers can have either single or split outputs. Figure 25(a) shows a single-output driver. In this case, a separate diode can be implemented to control turn-on and turn-off as explained earlier in section 2.9. It increases bill-of-materials cost taking more space on the gate driver board and adding impedance in the gate loop [29]. As an alternative, a split output driver has separated turn-on and turn-off paths for the complete control over drive source or sink strength [30]. Thus, the split output is the best option to control the power device efficiently and safely as shown in the Figure 25(b).

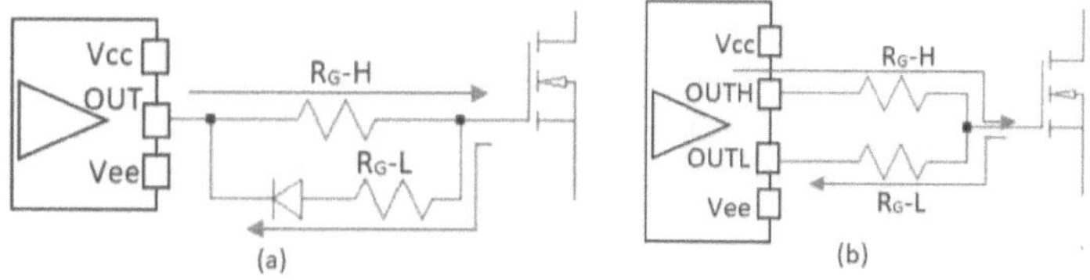

Figure 25 The gate driver solutions (a) Single channel output (b) Split outputs

The power dissipation splits between internal R_{Gint} and external R_{Gext} gate resistances. Therefore, attention must be paid to power ratings for the external gate resistance [26]. Different driver IC matching the target specifications are chosen for the final design as described below.

Table 2 Gate driver IC datasheet specifications

Gate drivers	Ucc27531	1ED020I12-F2	ADuM4135
Isolation type	External optocoupler	Galvanic	iCoupler
SCP protection	NO	Yes	Yes
Outputs	Split	Single	Split
Isolation voltage	$\pm10KV/\mu s$	$\pm100KV/\mu s$	$\pm150KV/\mu s$
$t_{d\text{-}H}/t_{d\text{-}L}$	17 ns	205 / 175	50 ns
t_r/t_f	15 / 7 ns	60 / 60 ns	16 ns
$I_{G\text{-}H}/I_{G\text{-}L}$	2.5 / 5 A	2.4 A	4.61 A

4.1.1 Gate driver A: Texas Instruments Incorporated - Ucc27531

The gate driver TI-Ucc27531 is named as gate driver (a) which is capable to drive the DUT with exceedingly small propagation delays. The input threshold of Ucc27531 is controlled by CMOS and TTL compatible low voltage logic. The driver is manifested with 1-V typical hysteresis feature offering better noise immunity. The Ucc27531 device features two outputs configuration for both turn-on and turn-off transients. The gate-drive current is derived via the OUTA pin and sink through the other OUTB pin. These split output pins arrangement let user to select two independent turn-on and turn-off gate resistors. Asymmetrical drive OUTA and OUTB easily control the switching gate driver up to 5-A sink and 2.5-A source peak current accordingly. The driver IC is based on totem pole configuration and ideally designed for switch mode power supplies. Ucc27531 gate driver includes the UVLO with hysteresis functionality [30]. When V_{dd} exceeds its threshold level the chip start to begin its normal operation. The chip holds the output low until threshold voltage has crossed.

Table 3 Specifications of gate driver (a) TI- Ucc27531

Driver (a) Texas Instruments Ucc27531	
Output Current	Peak 5-A sink and 2.5-A Source
V_{out} max	18V to 35V
Input Voltage	18V
Start-up current	$200\mu A$
V_{ON} supply (UVLO)	8.9V
PACKAGE	DVB

The driver IC doesn't come with an internal galvanic isolation thus needs an external isolation design. The selection of this gate driver makes possible to achieve very low propagation delay, high peak currents and negative biasing with minimum cost.

4.1.2 Gate driver B: Infineon- 1ED020I12-F2

The driver from Infineon-1ED020I12-F2 named here as gate driver (b) is a galvanic isolated single channel MOSFET gate driver which is available in DSO-16 package, providing the capability of single output current around 2A. The logic pins in the gate driver are CMOS compatible and can handle 5V inputs from the microcontroller. The best feature in Infineon-1ED020I12-F2 is the coreless transformer technology for galvanic isolations which gives an extraordinary protection required in high voltage applications. The best reason to choose 1ED020I12-F2 is the several protections features provided by gate driver IC like force shut down, desaturation protection, and active Miller clamping [31].

Table 4 Specifications of gate driver (b) Infineon-1ED020I12-F2

Driver (b) Infineon-1ED020I12-F2	
Output Current	2 A peak drive output
V_{out} max	30V
Input Voltage	12V to 28V
Start-up current	<100µA
V_{ON} supply (UVLO)	>11V
Package	PG-DSO-16-15

4.1.3 Gate driver C: Analog Devices- ADuM4135

The ADuM4315 (gate driver c) is a single-channel gate driver specifically designed for driving SiC MOSFETS with suitable protection features. The main aspect to choose the driver IC is extremely low propagation delay which can be achieved in chip scale transformer driver IC. The IC transformer is coupled with iCoupler technology capable of providing isolation between the output gate drive and input signal. The driver provides split outputs features which can be handy during high frequency and voltage applications. The driver IC provides isolated desaturation fault reporting and soft turn-off at shutdown. ADuM4135 is capable of provide peak gate current of 4A which can be very handy for fast switching devices like SiCs. The device starts operation when the UVLO level > 11V and both ready and fault pins are high. The chip also includes the DESAT detection circuit which provides enough protection against short circuit in IGBT/MOSFET operation [32]. This desaturation protection circuit provides masking time around 300ns to limit the noise in switching period and protects the device against high voltage short circuits.

Table 5 Specification of (c) Analog Devices- ADuM4135

Driver (c) Analog Devices-ADuM4135	
Output Current	4 A peak drive output
Outputs	Split Output Options
V_{out} max	30V
Input Voltage	12V to 30V
Start-up current	10μA
V_{ON} supply (UVLO)	11.1V
Package	PG-DSO-16

4.2 Power supply design

The power supply for the gate driver is one source of design differentiation. A simple solution can be an only positive supply to gate of the power MOSFET. But to mitigate the false turn-on effect in latest SiC devices a negative gate to source voltage is necessary. The simpler approach for this solution, consists of a positive supply sharing a negative voltage for turn-off with respect to source of the power MOSFET. The voltage level of gate driver power supply should be much greater than the threshold voltage of device [18]. The threshold voltage of the device normally depends on the doping levels and device materials [25]. The required positive voltage for a N channel device needs to match with the gate-source ratings of the DUT. There is a chance that due to high dv/dt caused by switching event that gate potential will rise due to effects of AC coupling of the C_{GD} gate drain capacitance. This ramp in the gate voltage will cause erroneous ON, thus a potential below than ground is necessary to avoid the false turn-on condition in the main application. The dI/dt relationship in previous studies shows that the higher gate to source voltage results in the lesser dI/dt, thus the V_{GS} should be close to high as much as possible [28]. Advanced gate driver ICs specially designed for SiC devices can be modified with asymmetric voltages levels. A two-level power supply is needed to match this condition. Therefore, isolated power supply from Murata exhibiting the required features shown in Table 6 is used.

Table 6 Specifications of DC-DC converter MGJ2D121505SC [33]

MGJ2D121505SC DC-DC converter for driver modules power supply	
Nominal Input Voltage	12V
Output Voltage 1	15V
Output Voltage 2	-5V
Output Current 1	80mA
Output Current 2	40mA
Input Current at Rated Load	120mA
Ripple & Noise (Typ)2	30 mVp-p
Efficiency (Typ)	80%
Isolation Capacitance	2.9pF

The MGJ2D121505SC can deliver 2W isolated power with a dual output of 15V/-5V. The negative sinking voltage is very helpful to turn the chip off slightly faster than 0V assuring that the device should not turn back on.

4.3 The gate driver schematic and prototyping

The important considerations are kept in mind while designing the layout for high frequency gate drivers for example parasitics can heavily influence the gate signals for SiC devices and can be deleterious for overall system [34]. The complete schematics design for gate driver (a) Ucc27531 chip from the topology shown below in the Figure 26. Extra decoupling capacitors are added to the power supply for a steady output needed for gate driver IC for minimizing noise and consuming all energy present.

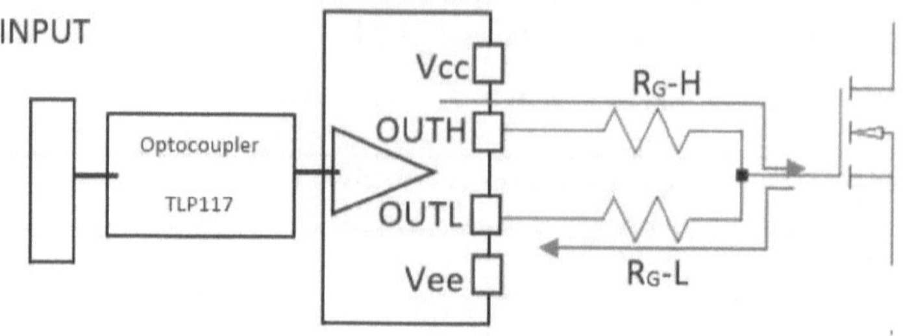

Figure 26 Gate driver Ucc27531 gate drive tropology

In addition, two gate resistors for Rg-on and Rg-off are added which help to maintain the switching speed of the DUT. Two hyper fast diodes with minimum recovery are added in the design to utilize the different turn-off resistance option in a single output driver like gate driver (b) 1ED020I12-F2. The gate driver boards also include a high accuracy and high PSRR voltage regulator for supplying the required power to opto-coupler and microcontroller for generating PWM signal needed for DPT test board. The IC is supplied by 5.2 KVDC isolated gate driver DC-DC converter, which provides the suitable supplies and isolation required for protection. The isolated DC-DC converter and voltage regulator schematic are shown in the Figure 27.

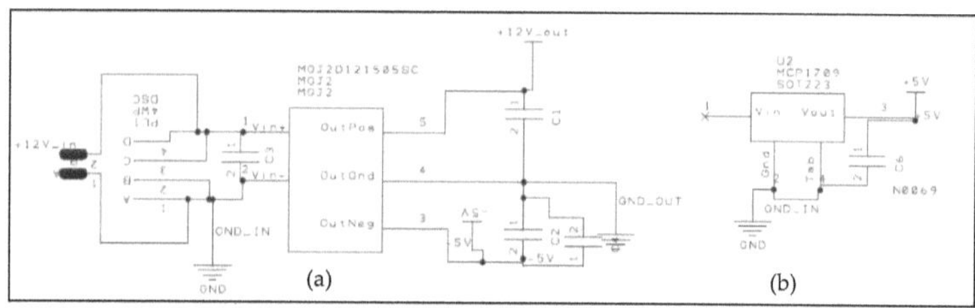

Figure 27 Schematics of (a) DC-DC converter (b) PSSR Voltage regulator

The PCB layout of gate driver is created keeping in mind the shorten traces for minimizing the parasitic inductances as much as possible. All the gate driver boards are kept at the same design layout to achieve similar parasitic in all the boards. The Figure 28-30 show respective gate driver application schematics.

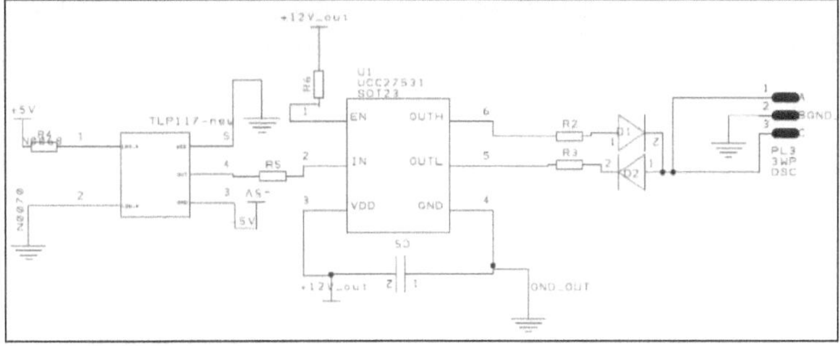

Figure 28 Schematic view design of gate driver (a) Ucc27531

Few jumpers are applied to the gate driver board to shorten the trace whereas its width is adjusted at a reasonable value for minimizing the parasitic inductances. The 30µF SMD capacitors are added in the design due to achieve higher transient drive.

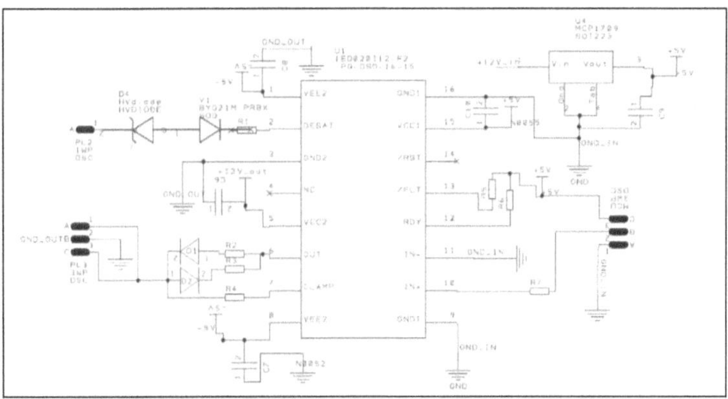

Figure 29 Schematic view design of gate driver (b) 1ED020I12-F2

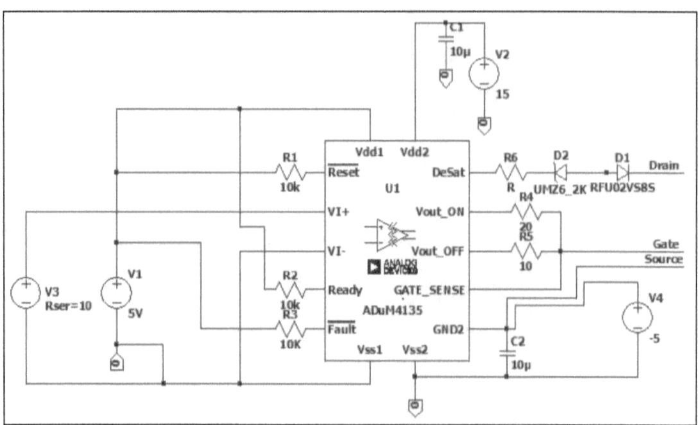

Figure 30 Schematic view design of gate driver (c) ADuM4135

The gate driver board is a two-layer board, and it is designed in a way on both sides for reducing noise in both layers [35]. In addition to low parasitic inductance in the gate driver loop, proper creepage and clearance distances are set in the PCB design for preventing insulation failures [36].

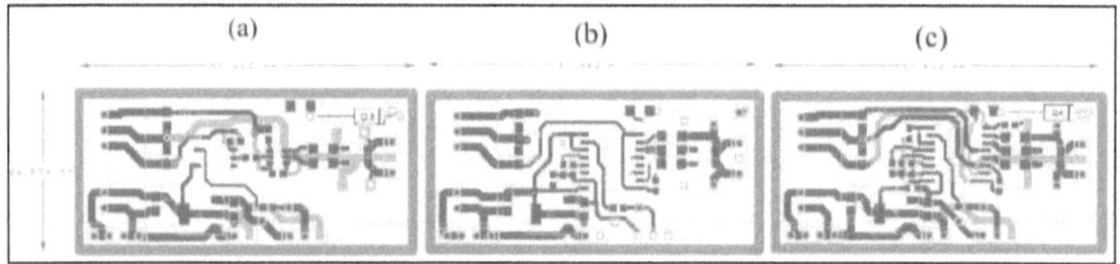

Figure 31 PCB layouts for the gate driver modules (a) Ucc27531 (b) 1ED020I12-F2 (c) ADuM4135

A high voltage diode is accounted for the desaturation protection. The gate driver (b) and (c) utilize this diode to block high voltage and measure the voltage drop across diode for DESAT protection features. Figure 31 describes the layout of the board while Figure 32 shows its real appearance.

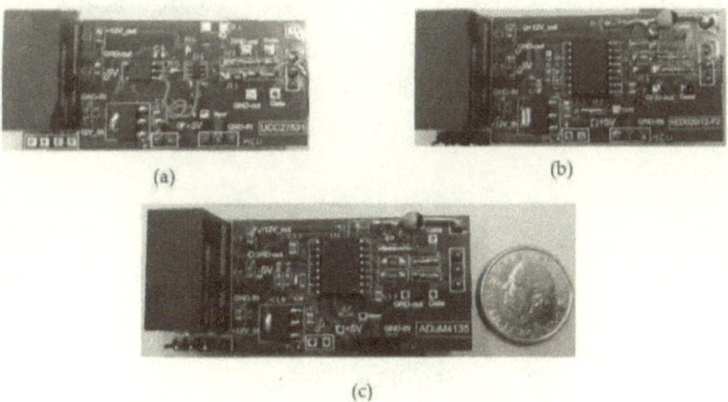

Figure 32 Real appearance of gate driver modules (a) Ucc27531 (b) 1ED020I12-F2 (c) ADuM4135

4.4 Double-Pulse Test Setup Description

The dynamic characterization platform as Double pulse tester is commonly used to evaluate the switching transients of SiC power discreates. A setup is designed which provides clean and accurate measurements under well-defined conditions which can efficiently correlate to the simulation results of the device. Important driver design parameters like maximum allowed leakage inductance and parasitic capacitance parameters can be derived from this setup [35]. The dynamic behaviour of the SiC MOSFET is characterized by voltage and current across the DUT during turn-on and turn-off switching under the defined circuit conditions. Consequently, characteristics like dv/dt, di/dt, and gate driver properties can be achieved from our DPT tester.

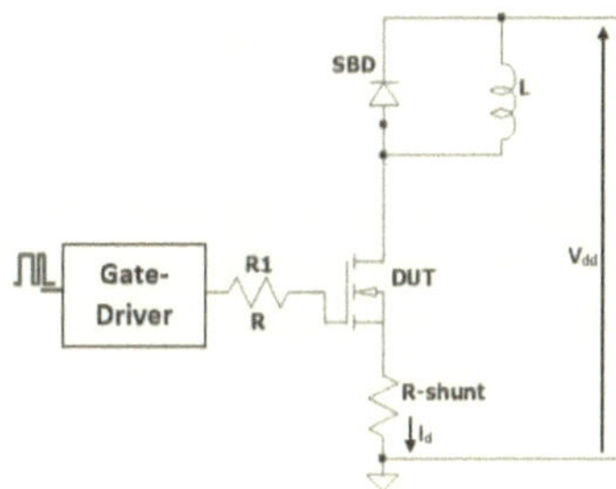

Figure 33 Double Pulse Setup

The tester is designed for DC-Link voltage up to 2KV and a current up to 50A. The PWM pulses for control part are sent by a designed microcontroller. The DPT can validate 1200V/40A SiC MOSFETs for experimental validation. The real appearance of DPT design setup is show below.

Figure 34 Developed Double Pulse test setup

4.4.1   Load Inductor

The specific load can be set by as an inductor and the pulse widths in the setup. The value of load inductor is calculated as follows:

$$L \geq \frac{Vdc}{\Delta I_L} = \frac{Vdc}{K\Delta_i \times I_L} t_{sw} \tag{1.16}$$

Where t_{sw} is the time taken by a complete switching event while V_{dc} and I_L are DC bus voltage and load current, respectively. K.Δi shows the current variation percentage during the switching transition ranging 1%–5%. Low limit of the required inductance is calculated in equation 1.18. If the inductance value is high the longer duration must be used for achieving the same current, which in turn causes thermal concern in MOSFET. The inductor used in the testing is a big 480μH inductor.

4.4.2   DC Capacitor bank

Energy is supplied by the two banks with possibility of series and parallel connection with large capacitance which establish the inductive current during the turn-on pulses. The minimum value for C_{bulk} is:

$$C_{bulk} = \frac{480\mu \times 24^2}{2 \times 0.03 \times 200} = 115\mu F \tag{1.17}$$

Our DC link board is available in two voltage ratings to cover two ranges of available rating for DUT. Four jumpers are placed to utilize different capacitance needed for different voltage levels.

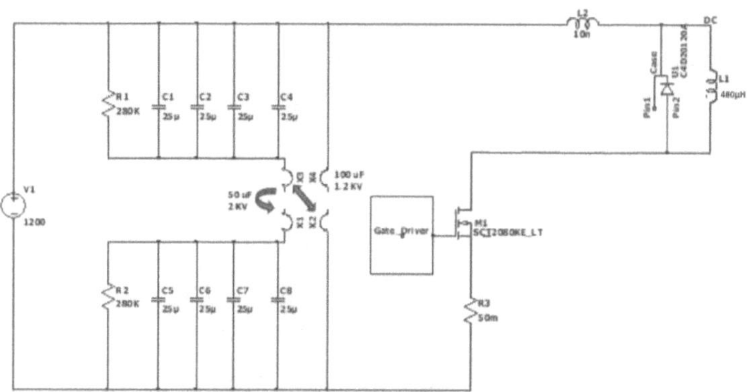

Figure 35 DC-link board capacitor schematic

By changing the different jumpers, two capacitance levels of 100µF for 1.2 KV and 50µF for 2 KV can be achieved.

4.4.3 Bleeder resistor

For safety reasons, two resistors are in parallel with the capacitor bank to discharge the capacitors under the condition when the test setup is turned off but also to balance the voltage across the capacitors. $R_{bleeder}$ is based on trade-off between the power loss and the discharge time.

$$P_{bleeder} = \frac{V_{dc}^2}{R} = \frac{800^2}{300 \times 10^3} = 4.8\ W \tag{1.18}$$

$$T_{discharge} = R_{bleeder} \times C_{bulk} = 300 \times 10^3 \times 50 \times 10^{-6} = 15s \tag{1.19}$$

In the final design, bleeder resistor consists of two 300 kΩ, 5W metal oxide film resistors in the series [35]. For voltage balance of the capacitor bank, these two resistors are also in parallel with series-connected energy storage capacitors individually as seen in the Figure 35.

4.4.4   The device board

The device board is designed for different TO packages and device board is mounted on DC-link board so that inevitable common source inductance can be minimized. For minimizing effective length of power devices, a gate driver board is placed on the device board directly in the way that two boards are connected with each other through the same connecter. Here, in the PCB the positive and negative planes are overlapping, with an isolation layer between them. Hence, the two planes overlapping thus giving a minimal parasitic inductance.

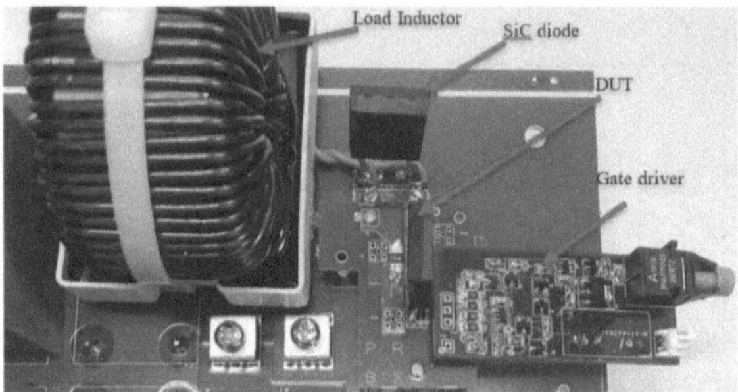

Figure 36 Device board placement

The parasitic minimization placement is based on minimization in the gate loop and power loop and components are placed as close as possible to each other [14]. RF connectors are placed just next to the device for voltage and currents measurements and a direct connection is made using a small coil thus resulting in minimization of group loop for oscilloscope. As a result, the lead lengths in both power and gate loop are reduced significantly. Parasitic minimization placement comparison is given in table below.

Table 7 Parasitic inductance based on different placements [14]

Placement style	Gate-loop induct-ance (nH)	Power-loop induct-ance(nH)	Common-source inductance (nH)
Conventional-placements	20.4	23.9	6.8
Parasitic minimi-zation placement	15.2	17.0	4.8

4.5 Measurement of voltage and currents

4.5.1 Oscilloscope selection

In a DPT, oscilloscope is being considered as the first measurement tool capturing the waveforms of the switching transients. The frequency depends on duration of the ramp and is independent to the voltage/current slew rate. For capturing the enough detail of a waveform, the bandwidth of the oscilloscope and the sampling rate should be higher than the equivalent frequency. WBG transients occur in the order of few nanoseconds. Thus, faster sampling rates and higher bandwidth improve the detail of the waveform. Tektronix MSO064 Mixed Signal oscilloscope is used for DPT measurements exhibiting features like sampling rate of 25 GS/s, and bandwidth rate of 8GHz to determine the accurately measured fastest switching transients. In addition, the oscilloscope provides the horizontal resolution to a maximum scale.

4.5.2 Voltage probe and current measurement

A 1GHz bandwidth voltage single-ended probe is selected which can measure voltage up to 300Vp-p. The voltage probe is manifested with low input capacitance specification (< 4 pF) for minimizing the loading effect of probe on the DPT circuit. Moreover, to obtain noise free measurements the ground path is shorted. Considering current measurement in DPT, inserting switching current measurement into switching loop is necessary, which affects the power-loop layout. In order to introduce the layout design and current measurement taken into account for the DPT, shunt resistance 50mΩ is added on low side for measuring the current in DPT setup [20]. The insertion inductance introduced by the shunt resistor is measured and was considered negligible.

5 Results and discussion

5.1 DPT test results

During the operation, two pulses are fed to the gate of SiC device for switching the device. The current though the inductor ramps up to required level during the first pulse before the DUT is turned-off. The turn-off transients are measured on the falling edge of first pulse (t_1). Meanwhile the turn-on trainsets are captured at rising edge of the second pulse (t_2). The switching voltage can be adjusted with the voltage V_{DC} and the current with the inductor charge time t_1. The implemented test conditions are illustrated in the Table 8 below. The typical logic signals observed in the double pulse signal is shown in the Figure 37, where -5V and +15V being low and high respectively relative to circuit ground.

Table 8 Parameters of double pulse test setup

Test conditions	Value
SiC Diode	C3D25170H
V_{DC}	300 V
I_d	15A
L	480 µH
C_{DC}	100 µF
$(t_0 - t_1)$	24 µs
$(t_1 - t_2)$	0.8 µs
$(t_2 - t_3)$	1 µs

Figure 37 Timing scheme for double pulse test

5.2   Selection of SiC DUT

The major benefits from the SiC MOSFETs over typical IGBT are reduced drive capacitances, the absence of tail current in SiC MOSFETs leads to faster turn-off and leads to better switching characteristics and energy loss. The information supplied from the manufacturers during the time of research, the energy during the turn-on and turn-off of SiC MOSFETs under research is depicted in the Table 9. The test conditions like gate resistance and gate drivers are not identical, a rough analysis performance for the devices is achieved from the data sheets supplied by manufacturers.

Table 9 Basic switching comparison of SiC MOSFETs

Device	Ciss (pF)	Coss (pF)	Turn-off Energies	Turn-on Energies
ROHM-SCT2080KE	2080	77	55 µJ	174 µJ
Cree-CMF20120	1915	120	155 µJ	125 µJ
Little-fuse 120E0080	1700	82	50 µJ	130 µJ

Alignment of the V_{DS} drain-source voltage and I_{DS} drain- source current and setup implementation are done in the same manner [34]. Because of the high-speed voltage transients of SiC, the switching transients result into high dV/dt as 100V/µs imposes common mode problems in the driver circuit. In addition, extreme dV/dt stress reduces the lifespan of the insulation material over time [37]. Different factors for the SiC power devices like rise time, fall time, and delays between turn-on and turn-off are usually different, and thus require individual consideration. The relationships between V_{GS}, V_{DS}, I_D and R_g are investigated in all the devices as shown in Figure 38-39.

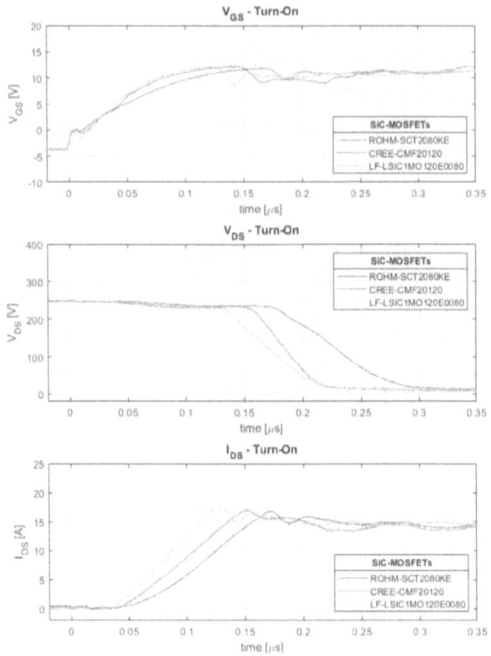

Figure 38 Comparison of different SiC Switches during Turn-on

The internal gate resistance of the SiC MOSFET ROHM-SCT2080KE, Cree-CMF20120, and LittleFuse-LFE0080 is 4Ω, 5Ω and 0.6Ω respectively, Due to lower internal resistance Rg-int the Little-fuse and Cree MOSFETs show the fastest rise and fall times but generate higher interference in EMI due to fast dI/dt. The output capacitances of the MOSFET also plays important role in EMI generation during transients. In SCT2080KE MOSFET V_{DS} shows better performance in turn-on transition. The SiC MOSFET from CREE-CMF20120 performs better in turn-off transition but generates higher ringing on the gate voltage.

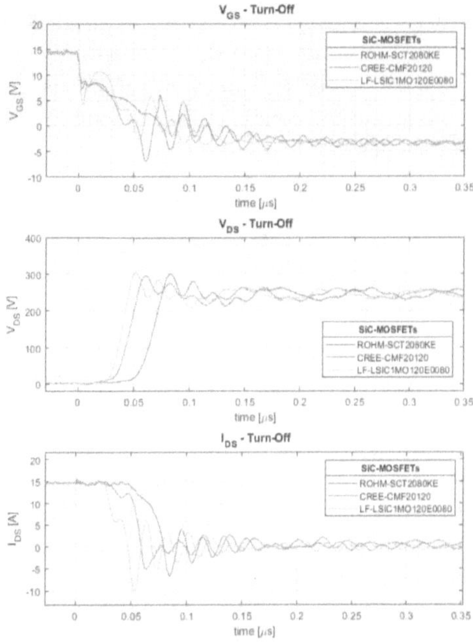

Figure 39 Comparison of different SiC switches Turn-Off

The SiC SCT2080KE switch has minimum threshold voltage 1.6V. The typical temperature coefficient of the threshold voltage is -5mV/C^0 [11]. Operation at an elevated junction temperature of 175 °C results in a threshold shift to roughly 1.6 V – (175–25) * 5 mV = 0.85 V. The output capacitance of the selected device is the minimum among other device i.e around 77pF. The ROHM SCT2080KE MOSFET improves the efficiency of overall system by delivering a steep rise and fall time that increases the application performance but limiting the excessive EMI.

5.3   Silicon carbide Schottky Diodes(SBDs)

The reverse recovery effect seen in the SiC switches heavily influence the switching losses in turn on period. The reverse recovery losses can be eliminated if the silicon freewheeling diodes are replaced with the latest schottky barrier diodes (SBDs) [38]. The SBDs exhibits very low reverse recovery charge Qrr due to its very low junction capacitance. SBDs are independent of forward current, temperature, and high dI/dt. The ultra-low Qrr specially helps in hard switch-based applications where the reverse recovery plays very important role. To verify the performance of advanced SBDs same switching test circuit (DPT) is used. The SBD C3D25170H is chosen for the double pulse test setup due to its superior dynamic performance as depicted in Table 10.

Table 10 SiC Schottky diode C3D25170H parameters

Manufacturer	Part ID	V_{RRM} (Peak reverse Voltage)	I_f (Forward current)	V_f (Forward voltage)	I_{RR} (Reverse current)	C (Total Capacitance)	Q_C (Total charge)
Cree	C3D25170H	1700V	26.3A	1.8V	20μA	187.5pF	230nC

For the 1.2kV device (SCT2080KE) testing, the slew rates for turn-off are set by 10Ω gate resistor around to 400V/μs and 500 A/μs. Meanwhile, the 20Ω is selected for turn-on period to achieve the di/dt around 250 A/μs. The waveform below shows the measurement of reverse recovery current at room temperature. Where the black represents the SiC MOSFET gate to source voltage, while the red, blue, and green show the current though diode, drain to source voltage, and voltage anode to cathode on SiC diode, respectively.

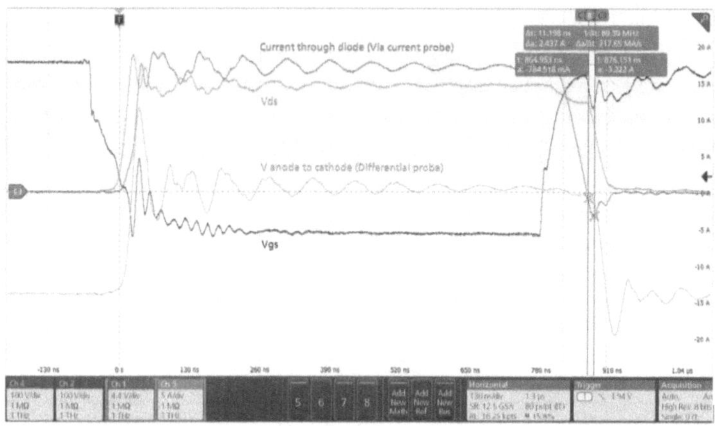

Figure 40 Reverse recovery time in SiC SBD diode C3D25170H

The experimental results show the peak reverse recovery current around 3.2A, which is approximate 80% reduction to normal pin diodes, and provides a recovery time around 12 ns, reduction of 90%, thus resulting in a peak instantaneous power of 0.5 kW, a reduction of 92% as shown in the Figure 41. The SiC SBD dissipates very small charge thus causing a very low voltage drop and reducing the overshoots results in drastic reduction in the switching losses.

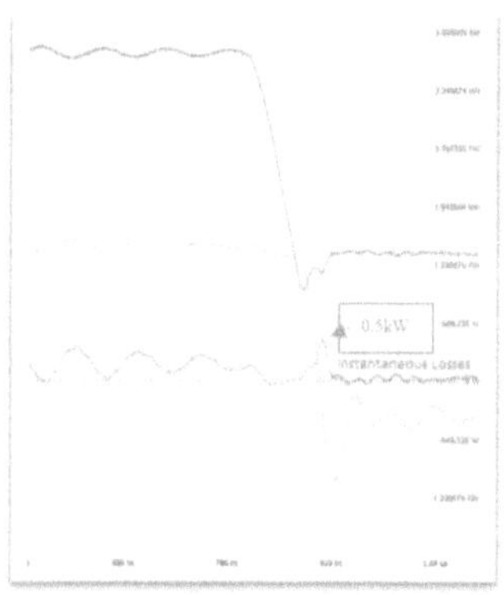

Figure 41 Instantaneous losses in SiC SBD C3D25170H

Here, the losses due to reverse recovery are the diode switching losses and the diode forward voltage causes the conduction loss. Due to around zero reverse recovery current in SiC SBDs, switching losses are almost negligible but the main power loss is the conduction loss due to forward voltage blocking [38]. However, in normal silicon diodes the conduction loss is less than the reverse recovery energy loss. So, the choice of much larger devices is very crucial to meet efficiency and thermal requirements. The Table 11 illustrates the overall improvement with SiC SBDs compared to conventional diodes.

Table 11 Improvement comparison in SiC SBD vs Si pin diodes

V_{cc} =300V I_D = 20A R_g = 10Ω				
Parameter	Unit	Si pin diodes	SiC SBD	Improvement (~%)
Peak reverse current	I_{pr} (A)	~13	5	88%
Reverse recovery time	T_{rr} (nS)	~82	12	148%
Recovery charge	Q_{rr} (nC)	~500	230	73%

5.4 The dV/dt and dI/dt

Different factors for the power devices like rise time, fall time, and delays between turn-on and turn-off are usually different, and thus require individual consideration. For example, the di/dt at turn-off can result into a large voltage overshoot, so it is beneficial to reduce the switching speed. This faster change in the drain voltage brings a negative effect to the MOSFET by charging the C_{GD}. Typically, in the MOSFET C_{GD} are much lesser than C_{GS} capacitance, which leads to the gate charge depletion and is proportional to the V_{DS}, dV/dt, and C_{GD}. The depletion in the gate charge results in slower gate charging and eventually making the rise time slower. Thus, the measurements are carried out on two different levels to observe the dV/dt and dI/dt phenomena. The voltage and current values ranging from 100V to 300V and 5-20A. V_{GS}, V_{DS} and I_D had voltage spikes and current spikes during both switching moments because of the parasitic inductances. As parasitic inductances resonate with the parasitic capacitance making the voltage and current oscillations [9]. Peak in gate to source V_{GS} is achieved at the end of the V_{DS} ramp. Few interesting points immediately demonstrated in Figure 42 that the high dI/dt caused by the rapid change in voltage can heavily influence the gate to source voltage. This change in V_{GS} can influence the switching dynamics. Thus, it is beneficial to study the influence of gate resistor for high dV/dt applications.

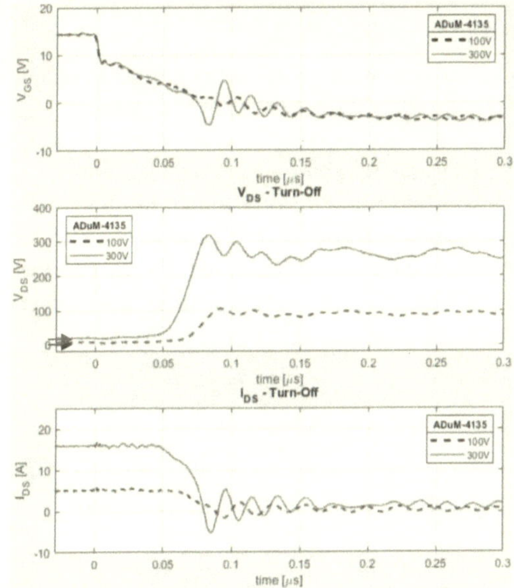

Figure 42 The effect of high dI/dt on the gate and drain voltage

It is also verified from the results the overshoots also change when changing the voltage and current magnitudes. Another interesting effect seen in the turn-off results corresponds to overshoot ringing frequency. The period of 100V signal is significantly different from the period at 300V seen in Figure 42. This period difference can be explained by the nonlinear capacitance behaviour of SiC MOSFET explained from the Figure 5 and can be expressed as:

$$f = \frac{1}{2\pi\sqrt{LC}} \tag{1.20}$$

$$L = \frac{1}{(2\pi f)^2 C_{oss}} \tag{1.21}$$

As the C_{oss} of SiC MOSFET varies significantly over the applied drain to source voltage. The ringing and overshoots are caused by the RLC circuit formed by the C_{DS} and C_{GS} from the MOSFET. The turn-on time shows small ringing as compared to turn-off and increase slightly with increased V_{DS}. The variation seen in turn-off is due to the output capacitance C_{oss}, where the load current is responsible for charging the C_{oss}.

5.5   Influence of gate resistance Rg

Gate resistors also control the speed of the transient voltage (dV/dt) and transient current (dI/dt) of the device. The presence of high gate resistance limits the noise in switching and but increases the switching losses. Moreover, the selection of resistor less than 5Ω results in significant gate voltage overshoots and bring it closer to the threshold voltage. It seems like 5Ω for turn-off is not correct and generates overshoots to the SiC MOSFET, therefore, the effect of Rg on the dynamic performance of the SiC MOSFET SCT2080KE, 10Ω to 30Ω with all gate drivers are further investigated.

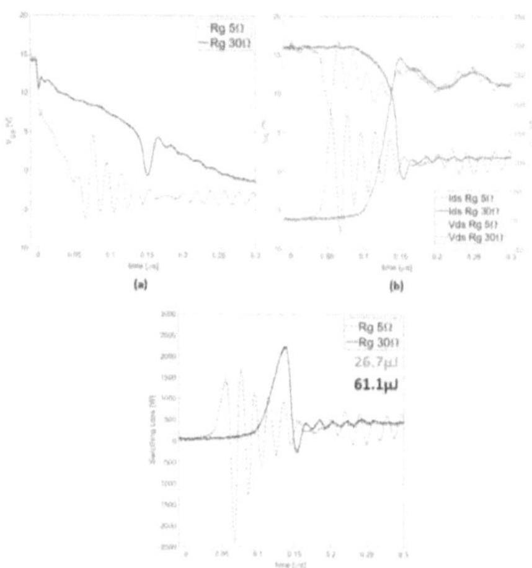

Figure 43 Comparison of gate resistor choice (a) Vgs (b) Vds and Ids (c) losses

Figure 44 shows the waveforms of V_{GS}, V_{DS} and I_D for all driver modules at the moment of SiC MOSFET turn-off and turn-on with different gate resistance values.

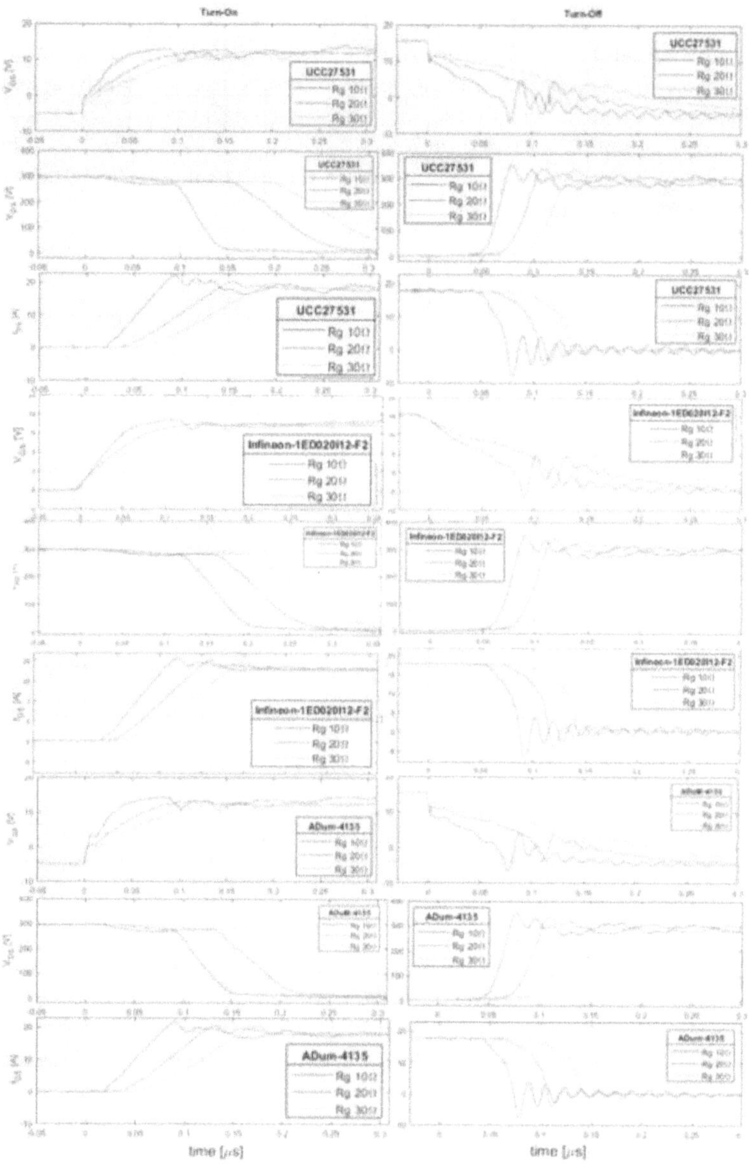

Figure 44 Turn-on and turn-off DPT results for gate drive (a),(b) and (c)

It is clear that the increase in gate resistance affects slew rate of the voltage rise and fall time. Both di/dt and dv/dt decrease gradually as the Rg increases. As an effect, switching speed decreases, and the switching loss increases. The dV/dt and dI/dt are taken from the Figure 44 as dV/dt (on) = (0.9 V_{DS} – 0.1 V_{DS} /tf) and dt/dt (on) = (0.9 V_{DS} – 0.1I_D/tr) as explained in section 2.5.1. The slew rates like dV/dt and dI/dt ratios for gate driver modules are summarized in Table 12-14.

Table 12 TI-Ucc27531 achieved parameters in DPT

R_g		t_d		V_{DS}		dv/dt		di/dt	
On	Off	Off	On	Fall	rise	On	Off	On	Off
10	10	53ns	80ns	19ns	58ns	4.02V/ns	13.87V/ns	316A/µs	684A/µs
20	20	73ns	105ns	26ns	85ns	2.46V/ns	9.14V/ns	190A/µs	396A/µs
30	30	99ns	279ns	31.6ns	118ns	2.0V/ns	7.27V/ns	152A/µs	418A/µs

Table 13 Achieved parameters in DPT for gate driver b (Infineon-1E0120I12-F2)

R_g		t_d		V_{DS}		dv/dt		di/dt	
On	Off	Off	On	Fall	Rise	On	Off	On	Off
10	10	80ns	120ns	23ns	71ns	3.31V/ns	11.58V/ns	255A/µs	526A/µs
20	20	80ns	166ns	27ns	98ns	2.43V/ns	8.71V/ns	192A/µs	373A/µs
30	30	112ns	ns	34ns	124ns	1.90V/ns	6.82V/ns	148A/µs	369A/µs

Gate resistance influences the rise and fall times of gate to source voltage V_{GS} and switching times are much reduced by providing a higher dV/dt and dI/dt values. In trade off of faster turn-off but lesser overshoots a slightly bigger resistance of Rg-off 10Ω is chosen. However, turn-on can be further slowed down by using a gate resistor value of 20Ω. This not only reduces the overshoots in the switching but will also slow down the reverse recovery effect thus producing the trade-off between faster transition and lower reverse recovery effect.

Table 14 Achieved parameters in DPT for gate driver c (ADuM4135)

R_g		t_d		V_{DS}		dv/dt		di/dt	
On	Off	Off	On	Fall	rise	on	off	on	off
10	10	49ns	81ns	19.7ns	65ns	3.67V/ns	12V/ns	311A/μs	577A/μs
20	20	74ns	135ns	26ns	74ns	3.19V/ns	8.81V/ns	228A/μs	445A/μs
30	30	93ns	180ns	31ns	97ns	2.41V/ns	7.20V/ns	178A/μS	384A/μs

5.6   Gate driver results

The gate drivers are further characterized on the DPT platform and the waveforms comparison with all the gate drivers are shown in the Figure 45. The results were characterized on the same Rg-off and Rg-on values of 10Ω. A more prominent delay can be also observed within the gate drive (b). As the slow rise and fall time can influences ultimately the switching characteristics by generating more losses. These slower transitions cause the less severe over-voltage and ringing during the turn-on and turn-off as compared to gate driver (a) and (c).

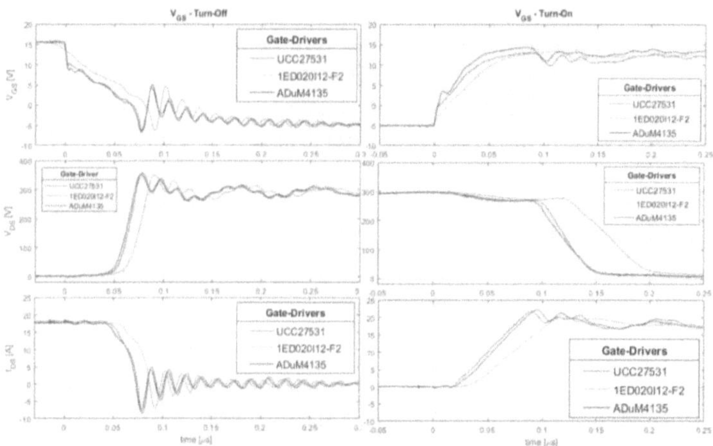

Figure 45 DPT results with all gate drivers

Table 15 Gate driver (GD) results (a) Ucc27531 (b) 1E0120I12-F2 (c) ADuM4135

GD	t_d		V_{DS}		Slew rates dv/dt		Slew rates di/dt	
	Off	On	Fall	rise	on	off	on	off
(a)	53ns	80ns	19ns	58ns	4.02V/ns	13.87V/ns	316A/μs	684A/μs
(b)	80ns	120ns	23ns	71ns	3.31V/ns	11.58V/ns	255A/μs	526A/μs
(c)	49ns	81ns	19.7ns	65ns	3.67V/ns	12V/ns	311A/μs	577A/μs

The shorter rise and fall times in V_{DS} by the SiC MOSFETs cause more overshoots in the waveforms with the lower resistance values. The ADuM4135 somewhat provides the lesser under/overshoots in dI/dt while providing faster ramps and less noise in switching transitions. The overshoots wise the driver (a) and (c) performs much similar.

5.7   The gate Current

The gate driver must be able to supply maximum current during the region require to reduce switching losses. This is dependent on the gate resistor and drive voltage during that plateau. Charging the gate capacitance as fast as possible is needed to minimize the switching time [39]. Hence, a driver with maximum current capability should be used. The maximum current is a characteristic that can be taken from the datasheet of the IC. That characteristic needed to follow the specification, as given in following equation:

$$I_{Drive} = \frac{Toral\ charge}{Commutaion\ time} = \frac{Q}{t_{d(on)} + t_r} \tag{1.22}$$

Where the commutation time is the sum of $td_{(on)}$ and rise time. Here, the gate driver IC should be able to provide the required voltage levels and current level for the SiC MOSFETs. This peak current rating is rarely, if ever observed in practice as the pull-up and pull-down networks gradually ramp up to full strength. The calculated required totem pole current for different SiC MOSFETs are shown in the table.

Table 16 Required current rating for SiC devices

Manufacture	Model	Required totem Pole Current Rating
ROHM	SCT2080KE	1.49A
Infineon	IMZ120R045M1	1.92A
Little Fuse	LSIC1MO120E0080	4.75A
ST	SCT20N120	1.73A
Cree	C2M0080120D	2.12A

In our previous studies showed that the maximum current can be achieved on selection of low gate resistance value of Rg. Selection of higher gate resistance generates high switching losses but results in less amount of EMI during switching transition.

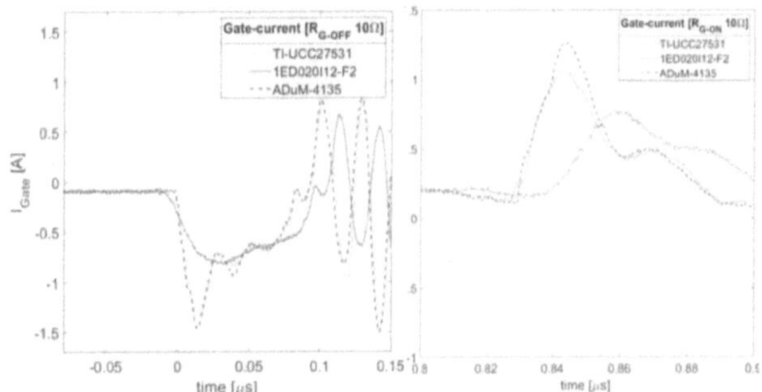

Figure 46 Experimental gate current for the period turn-off(left) and turn-on(right)

5.8   DPT Switching losses

The instantaneous switching losses are calculated as explained in the section 2.8. The observed losses in DPT experiment with all the gate drivers are created with MATLAB. The higher current capability results in lesser switching losses. The losses for 1ED020I12-F2 show higher switching losses as compared to other drivers due to high slew rates during switching intervals. The Ucc27531 shows lowest switching losses during turn-off period. But for turn-on period the same driver ranks to 2nd lowest in losses. As ADuM4135 crosses switching interval quickly and provides highest current results in lessen switching losses.

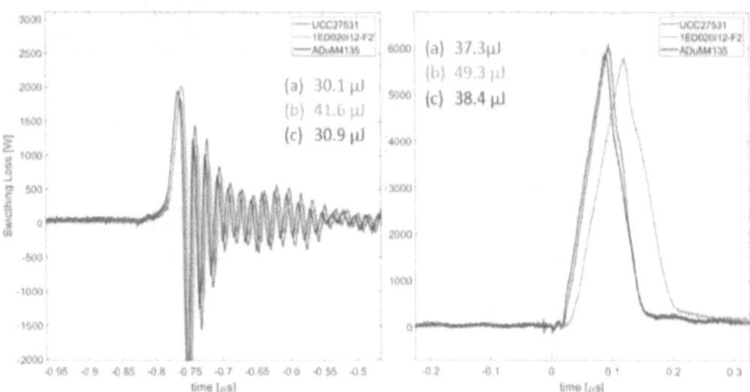

Figure 47 Switching energy losses (a) Turn-off (b) Turn-on

5.9   Propagation delay

This delay affects the timing of the switching between devices, which is critical in high-frequency applications where the dead time is necessary so that two devices do not turn on at the same time, which would cause shoot-through and catastrophic failure [39]. If the dead time is smaller than the propagation delay, then both devices will turn on at the same time, as shown in Figure 48.

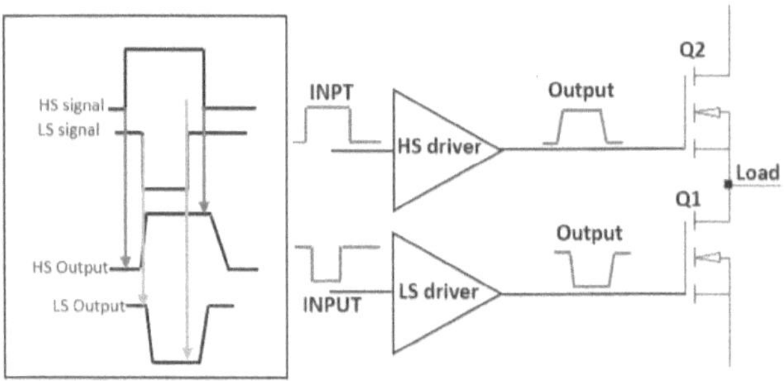

Figure 48 Mismatch in propagation delay

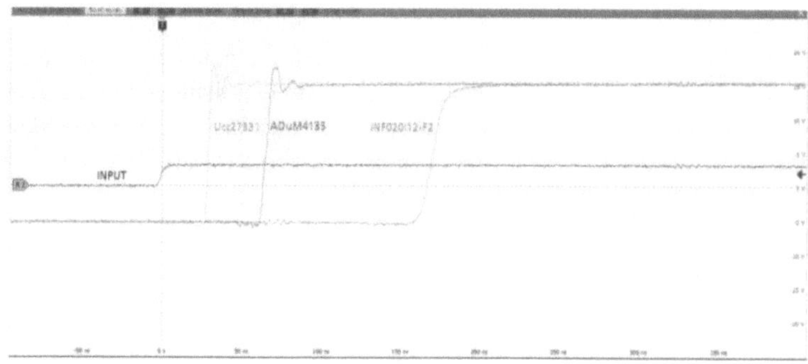

Figure 49 Turn-on propagation delays in gate driver a (Ucc27531), b (1ED020I12-F2), c (ADuM4135)

Table 17 Propagation delay measurements in gate drivers

Gate Driver	Turn-on delay	Turn-off delay
a (UCC27531)	30.745ns	26.10ns
b (1ED020I12-F2)	164.3ns	143ns
c (ADuM4135)	63ns	53ns

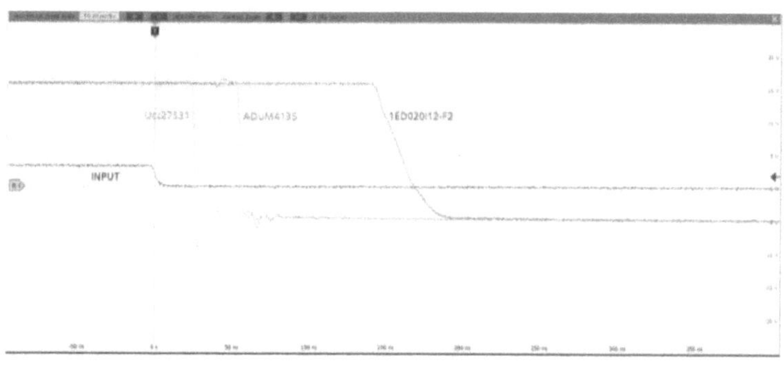

Figure 50 Turn-off propagation delays in gate driver a (Ucc27531), b (1ED020I12-F2), c (ADuM4135)

It is best to use the gate driver with low pulse width distortion, small rise, and fall times, and low propagation delay. All the gate drivers are triggered at the same time, least propagation delay was achieved in the gate driver (a) Ucc27531 due to external optocoupler which has few ns delay internally. The worst propagation delay is seen in the gate driver (b) which is due to internal transformer used in the gate driver IC.

5.10 Short circuit protection

The short circuit protection present in the gate driver modules (1ED020I12-F2) and (ADuM4135) is tested in this section. The driver ICs comes with built in voltage comparator around 9V. The fault condition can be implemented by using multiple DESAT series diodes as shown in equation 1.25.

$$V_{DESAT} = 9V - n \times V_f \tag{1.23}$$

Where Vf is the diodes forward voltage, and n is the number of diodes. To have a delay in fault triggering a C_{DESAT} can be used as:

$$C_{DESAT} = \frac{t_{blank} \times I_{CHG}}{9V} \tag{1.24}$$

The short circuit protection test is conducted with a turn-off resistance Rgoff = 10 Ω for both gate drivers. The test of SCP test is verified on designed double pulse tester at 50V drain to source voltage with C_{blank} of 0.1μF capacitor. A drain to source voltage of $V_{DS},scp = 2.4V$ corresponds to activation of SCP in the driver IC. This is achieved with a Zener diode of 6.3V as required by driver IC. The placement of SCP DESAT with the driver IC is shown in the Figure 51.

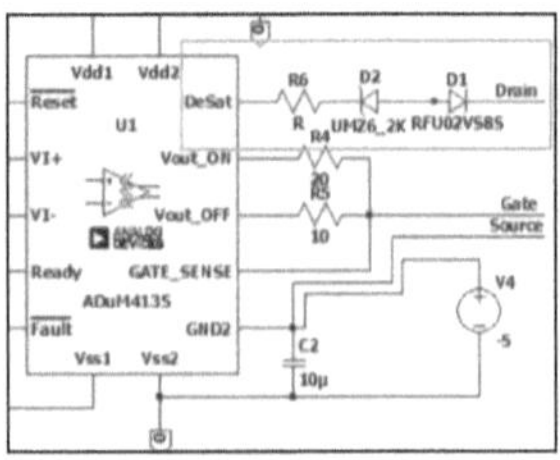

Figure 51 The DESAT protection in ADuM4135 driver module

Here, a drain to source voltage of $V_{DS,scp} = 2.4V$ corresponds to drain current of Id,scp 30A with R_{dson} of 80mΩ.

$$Id_{SCP} = \frac{2.4}{80m\Omega} = 30A \tag{1.25}$$

Figure 52 shows the test results of SCP at driver modules B (1ED020I12-F2). In the figure the grey waveform is the signal fed to gate driver IC where black, blue and yellow represents the gate to source (V_{GS}), drain to source (V_{DS})and the drain current Id. With a 12µs first pulse from microcontroller the SCP turn on can be seen around ~7 µs and 28A.

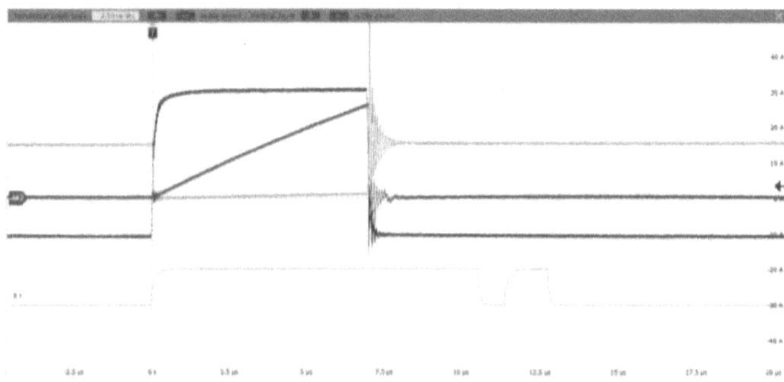

Figure 52 1ED020I12-F2 short circuit protection test

But the SCP turn-off event occurred only in 14ns which is quite faster thus producing very high dV/dt observed in the waveform. The reason is the driver IC from Infineon the SCP protection pull down resistance is very lower (not mentioned in the data sheet). While in the Figure 53 presents, the SCP result from gate driver (c) (ADuM4135). The SCP turn-off event in gate driver (c) ADuM4135 is observed to be slower i.e 62ns due to 300ns masking time provided by the DE-SAT pin thus producing a softer turn-off. If there is no SCP observed during the switching train sets the fault pin is low within the driver IC. As long as the SCP event has occurred the fault pin outputs 5V and pulls the gate voltage to low with a high resistor with driver IC.

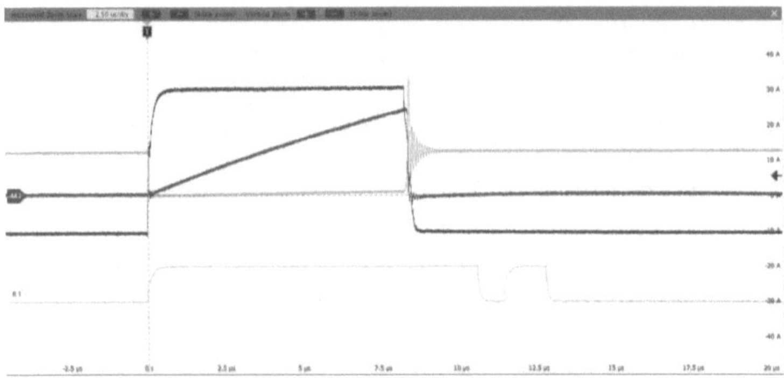

Figure 53 Gate driver ADuM4135 short circuit protection test

As the SCP in gate driver module needs to force a safe turn off at higher (Id) drain current to slow down the turn-off in order not to cause the high switching stress on MOSFET thus ADuM4135 conduct a safer turn-off switching with a masking time 300ns.

5.11 Simulation verification

Although, the double pulse test produces a fine solution for switching characteristics but, there can be still accuracy problems in measurements. As in the circuit loop presents the high dV/dt and dI/dt, the overshoots in measurements can be higher. As a result, the small capacitances and inductance in the loops can affect the accuracy of measurements for V_{DS} and Id. Thus, a simulation verification has performed keeping in mind with the approximately same power loop, gate loop and common source inductances. Different simulations were carried out to verify the feasibility and validity of our test setup. The simulations were mainly done on the gate driver (c) ADuM4135 using the software LTSPICE.The gate to source voltage V_{GS} and drain to source current I_D matches with the simulation results. It seems that the few ns delay in drain to source voltage V_{DS} is caused by the simulation model of the device.

Table 18 DPT simulation parameters

Parameter	Value
SiC MOSFET	SCT2080KE
SiC Diode	C3D25170H
V_{DC}	300V
I_{DS}	~20A
L	~480µH
C_{Dc}	100µF
Rg-on	10Ω
Rg-off	10Ω
Power loop inductance	20nH
Gate loop inductance	15nH
Common source	5nH

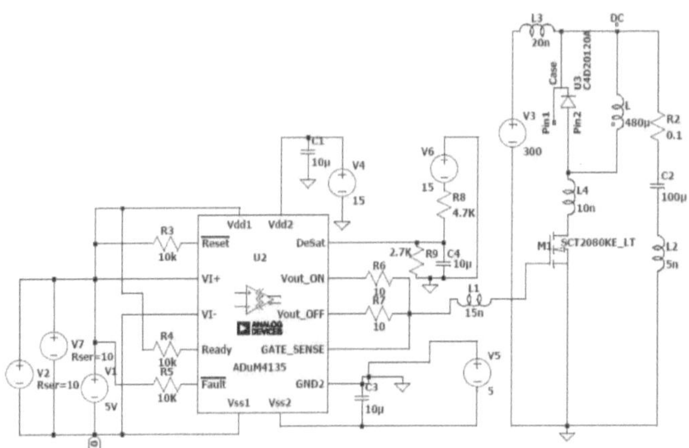

Figure 54 DPT LT-spice Simulation circuit for ADuM4135

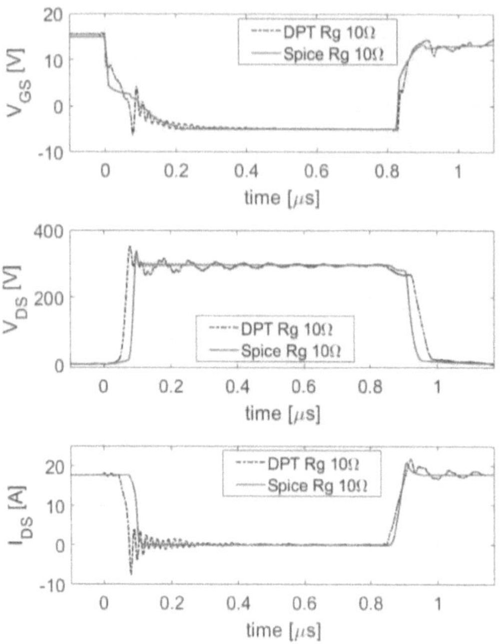

Figure 55 Experimental and LT-spice simulation results for module (c) ADuM4135

The simulations show that the turn-off slew rates matches with the experimental validation but in the turn-on event, different slew rate and very few nano second delay was observed. Even though with a very simple circuit, DPT design must be very high requirement for the measurements and PCB design layout design. In DPT design, both the main loop and gate charge loop traces are wider and minimized for a lower stray inductance. Thus, the measurements showed less oscillations and overshoots and were in close range with the simulation results.

6 Prototype of 1kW SiC based Half-Bridge DC-DC converter

To give an experimental verification of the designed gate drivers and theoretical expectations a prototype SiC based half bridge DC-DC converter is realised, which provides 110V and ~1kW output power levels.

6.1 Background

The major advantages in using a switch mode isolated power supply include having a higher efficiency and lower losses in the converter which ultimately result in small and less complicated cooling solutions [40]. Moreover, a significant protection can be obtained due to isolated transformer used in the power supplies. Usually, manufacturers use half or full-bridge configurations to increase the life span of the power supply as this topology produces the less stress on the switches [41].

6.1.1 Half-bridge DC-DC Converter

Half bridge topology is generally used for high stress applications greater than 200W. The Voltage stress on FET is the input voltage present for supply but only half of the input voltage is present across the transformer winding. Two switches at primary side Q1 and Q2 are in H-bridge configuration and each switch operates in complementary mode with suitable dead times and are controlled with designed gate drivers. The capacitors C1 and C2 possess same value and works as voltage divider and an DC bus capacitor. The transformer secondary can be arranged using a center tapped transformer configuration, where the center pin connects the LC filters output.

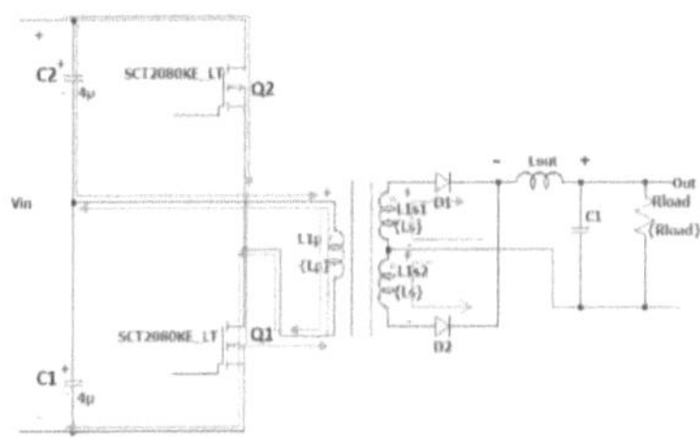

Figure 56 Stage 2 in half-bridge converter

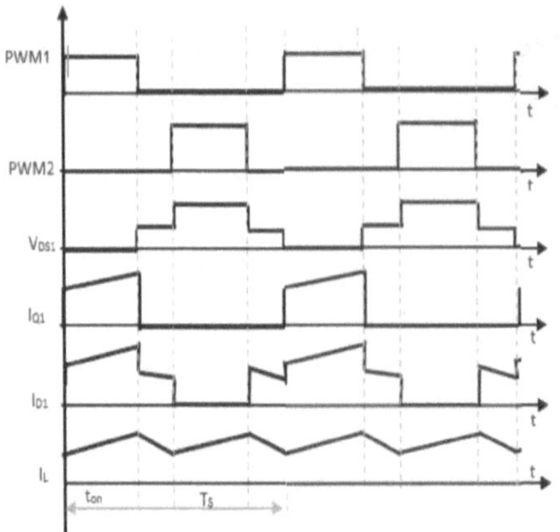

Figure 57 Typical Switching waveforms for Half-bridge converter

The half-bridge topology presents no excessive ringing at switch node and body diodes clamps the leakage energy and recovers to the input capacitors thus resulting in low EMI and higher efficiencies [40]. Whereas the capacitive divider balances the flux inside the transformer. The Figure 57 shows the operational stages for a half bridge converter. Firstly, the primary upper MOSFET Q2 is on, the voltage across the transformer primary is:

$$V_P = \frac{1}{2}V_{in} \tag{1.25}$$

Whereas the N_s / N_P is the turn ratio. The voltage at the transformer secondary is

$$V_s = \frac{1}{2}V_{in}\frac{N_s}{N_p} \tag{1.26}$$

This is first power stage and power is being transferred to the secondary output side. Here, the rectifier diode D1 is conducting and supplying the power to the output filter. In the second stage the Q1 is on and voltage across primary side and secondary are:

$$V_p = -\frac{1}{2}V_{in} \tag{1.27}$$

$$V_s = -\frac{1}{2}\left(V_{in}\frac{N_s}{N_p}\right) \tag{1.28}$$

In this stage the diode rectifier D2 is conduction mode and the secondary diode voltages are:

$$V_{D2} = \frac{V_{in}}{2}.\frac{N_s}{N_p}$$

(1.29)

The MOSFET Q1 should be rated as V_{DS1} = Vin. The capacitor divider also helps in balancing the flux at the transformer. The peak current IQ and the current through secondary diodes are:

$$I_{Q1} = \frac{N_s}{N_p}I_o\frac{\Delta I_L}{2}$$

(1.30)

$$I_{D1} = I_{D1} = I_o + \frac{\Delta I_L}{2}$$

(1.31)

During the free-wheel state where both primary switches Q1 and Q2 are off, the current I_L at secondary continues to flow. Here the total current is the inductor current. During the off time magnetizing energy forces the demagnetizing current across secondary winding and each secondary has half of the current, so there is no more flux at the core to release.

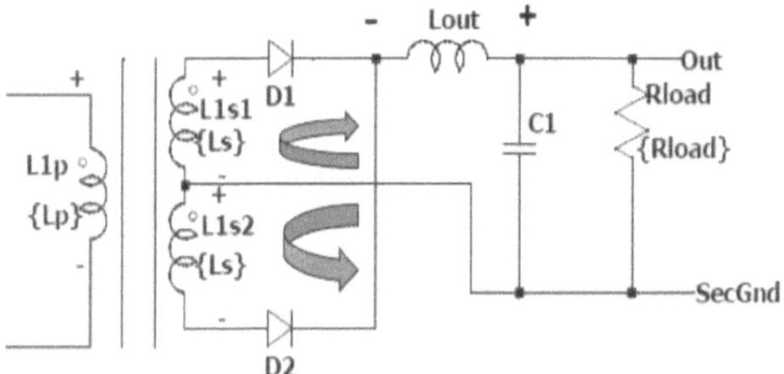

Figure 58 Secondary side voltages

In the off state both D1 and D2 are conducting and carrying half of the inductor current. The duty cycle for each primary switch is:

$$D_{Q1} = D_{Q2} = \frac{V_o}{V_{in}}.\frac{N_s}{N_p}$$

(1.32)

This results into output relation given as:

$$V_o = \frac{N_s}{N_p}.D.\frac{V_{in}}{2}$$

(1.33)

66

The main advantage of H-bridge converter is the maximum effective duty cycle of the topology which can approach near 100% on the low input voltage conditions. This results into a smaller input and output filter as well as smaller transformer design.

6.2 Design and dimensioning of converter

During the last phase of the SiC MOSFET module experimental tests, the half-bridge converter is built and tested. The idea is to obtain voltage and current parameters close to nominal without dissipation of significant amount of the power. In such a circuit energy is circulating between DC-side capacitors and inductor. Consequently, HV supply feeds only power losses in the circuit mainly in semiconductors and inductors. The SiC MOSFETs are controlled with an external microcontroller in complementary mode. The switching frequency is selected arbitrary about 100kHz. An input filter is added to protect the power supply for surge and EMC protection. The half-bridge converter is controlled by means of input voltage, switching frequency, and load current to reach various power levels.

Table 19 Object specification of DC-DC converter

Energy flow	Unidirectional HV-DC to LV-DC
Efficiency	95% peak
Input voltage	500-750V
Output voltage	105-115V
Maxi output power	1KW
Mini output power	0W
Maxi continuous output current	9A
Operating temperature	-30ºC to 70ºC

6.2.1 Transformer selection

The transformer in this topology can be designed with fine utilization if winding space helps in suitable isolation required for application. The selected transformer is selected from POWERBOX AB due to limitations and need. For the sake of half bridge configuration the transformer used in the converter is center taped for secondary winding. The selected transformer has 36 turns at primary and 26 turns for the secondary side. Where primary inductance and secondary inductance are measured around 1.3mH and 678µH, respectively. The saturation frequency is measured around 120 kHz.

Table 20 Selected transformer specification

Topology	Half-Bridge
Nprim	18 + 18 turns
Nsec	26 turns
P1 + P2 inductance	1,3 mH
S1 + S2	678 µH
Input voltage	350 / 1000 Vdc
Rated Power	500W
Rated working frequency	100 kHz
Insulation voltage	8 KVrms

6.2.2 SiC MOSFETs selection for prototype converter

The SiC MOSFET should be selected as per the break-down and its Rds(on) value. High voltage SiC MOSFETs are considered due to their superior dynamic characteristics as mentioned in previous section 5.2. The parameters for selected device are given in Table 21. Usually, converters under hard switching applications require a power MOSFET with minimum reverse recovery current. The longer reverse recovery effect can cause shoot through between the MOSFETs. The SiC SCT2080KE offer the lowest reverse recovery charge Irr, low total gate charge Q_g, thus suitable for the hard switching application and frequency range around 100kHz.

Table 21 Selected SiC MOSFET for prototype converter

Manufacturer	Rated voltage	I_d	$R_{ds(on)}$	Ciss	Coss	Q_g	Irr	Qrr
ROHM(2nd gen) SCT2080KE	1200V	40A	80mΩ	2080 pF	77 pF	106 nC	2.3 A	44n C

The selected 2nd generation MOSFET SCT2080KE can deliver the suitable rise and fall time by limiting the dV/dt and dI/dt required for our application. The device has the minimum RDS-on value to limit the conduction losses. This SiC MOSFET offers low input capacitance Ciss value of 2080pF which is helpful to minimize the rise and fall times. It also offers low reverse recovery current (Irr) of 2.3A which can be beneficial during hard switching. Altogether, all these parameters greatly affect switching speed.

6.2.3 Output Rectifier Stage

The centre-tapped full wave rectifier utilized the two secondary windings of a transformer in a way that only two diodes are needed in the rectifier. The number of turns for both windings on the secondary side as well as the voltage are same across each winding. When V$_{rectifier}$ is positive, D1 will be forward biased while D2 is blocking and the current flows out from the dot on the first winding on the secondary side. When the input voltage changes polarity, D2 will be forward biased while D1 is blocking, and the current flows into the dot of the second winding on the secondary side [42]. Since this centre-tapped half wave rectifier uses only two diodes, the voltage drops and conducting losses are lower for the diode bridge, making it more suitable for high voltage and low current systems.

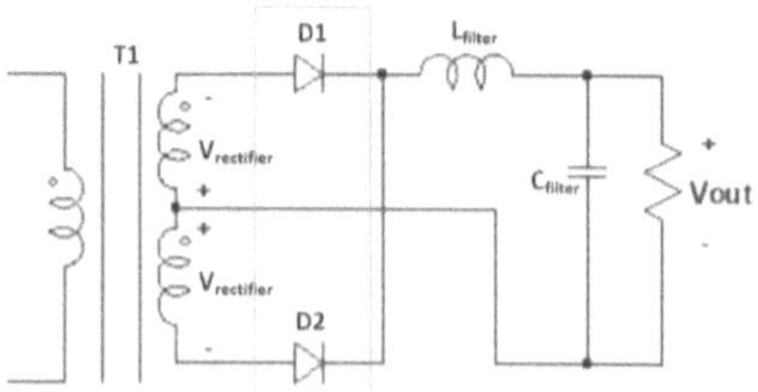

Figure 59 Rectifier at secondary side

$$V_{D1} = V_{D2} = \frac{V_{in}}{2} \cdot \frac{N_s}{N_p} \tag{1.34}$$

$$Peak\ Current = \ I_{D1} = I_{D2} = I_o + \frac{\Delta I_L}{2} \tag{1.35}$$

Table 22 Diode specification at secondary side

Model	DPG10I400PM
Package	TO220
Voltage ratings	400V
Current ratings	10A
Reverse recovery	45ns

6.2.4 Output filter at Secondary side

The half-bridge rectifier generates high ripples in the voltage and current. Therefore, to get a smooth output current and voltage, it is important to design an appropriate output filter for reducing the generated ripples [38].

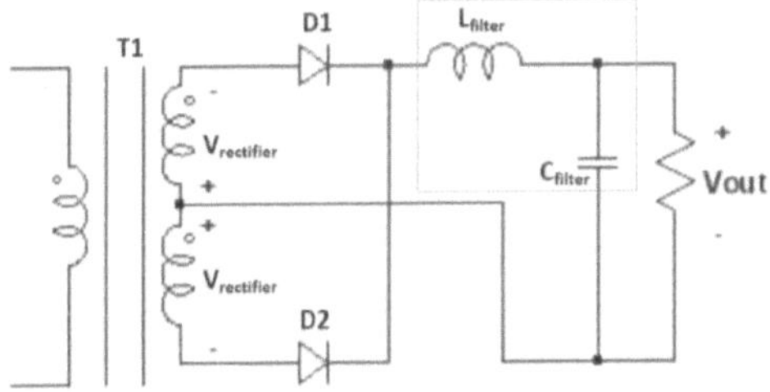

Figure 60 First stage Output filter at secondary side

The current in inductor from zero to DTs is considered. The voltage over inductor is:

$$V_L = L \frac{di_L}{dt} \tag{1.36}$$

$$L = \frac{V_L}{\frac{di_L}{dt}} \tag{1.37}$$

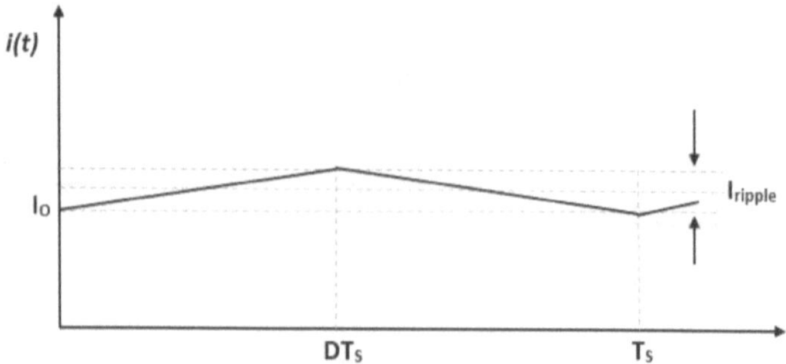

Figure 61 LC filter waveform

A minimum value of the inductance can be calculated when the voltage across the inductor reaches its maximum VL,max. The lowest necessary value of the inductance in order to limit the current ripple is calculated as shown in the equation below. The value of I_ripple is selected around 40% of I_out.

$$L_o = \frac{nV_i - V_o}{I_{ripple} \times I_o}(DT_{sw})$$ (1.38)

$$C_o = \frac{\left(\frac{1}{2 \times f_{sw}}\right) I_{ripple} \times I_o}{2 \times V_{ripple} \times V_o}$$ (1.39)

Where Lo is output inductance, n is transformer ratio, Vo is output voltage, Io is output current, Vi is input voltage, D is duty cycle, Iripple is current ripple and fsw is switching frequency. From above equations, the calculated required inductance value is ~700uH. The core used for the inductance is Shulin BS-1026, supplied from PowerBox AB. The calculated number of turns are around ~70. The other component of the output filter is the capacitor. The value of the capacitor should be high enough to keep the output voltage ripples within the limits. The output capacitor is selected around 200V rating one.

6.2.5 Second stage snubber design

A typical snubber circuit is used along with each diode at rectifier stage to suppress noise at secondary stage as shown in the Figure 62. The snubber capacitance should be higher than the resonant circuit capacitance, but it should be small enough to keep power dissipation in the resistor to minimum. The value for the snubber resistor should be close enough to the parasitic resonance impedance as mention in the following equation 1.41.

$$R_{sn} = \frac{3T}{2\pi \times C_{sn}}$$

(1.40)

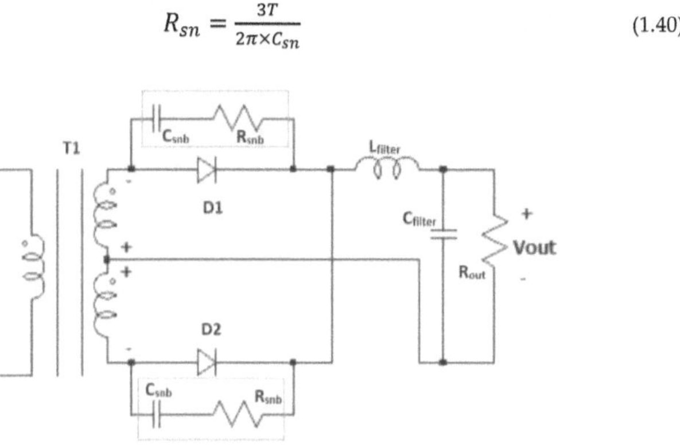

Figure 62 Snubber circuit at secondary side

6.2.6   PCB layout

To simplify the complexity, two SiC MOSFETs connected in half bridge configuration and controlled by the gate driver modules are implemented. Input and output filter are added to design for inrush and surge protection as the protype is operating at higher voltages like 500V input. Few high voltage ceramic capacitors are also added for the high voltage supply to lower the loop inductance. Printed circuit boards (PCBs) are more importantly designed to minimize the wire lengths and reduce the stray inductance and EMI for the design [13]. The high power and ground planes in the design are placed in upper and middle layers, respectively. while signal and signal ground in 2nd and 3rd layers. To power the ICs, the short and wider traces are placed as recommend for the PCB design [38]. The printed circuit board design likely requires different nets for carrying different current ranges. The basic need is to provide wider traces if it carries more than 0.3A. Thus, for main power the traces are wider and more shorted in the final design. As in switch mode power supplies, the isolation becomes a great challenge, thus the final design provides a suitable creepage distance between high and low voltage traces. Most important consideration is to route traces carrying the rapidly changing currents with shorter and wider lengths resulting in the decrement of di/dt. The basic schematic and PCB layout with four layers is constructed as shown in the Figure 63.

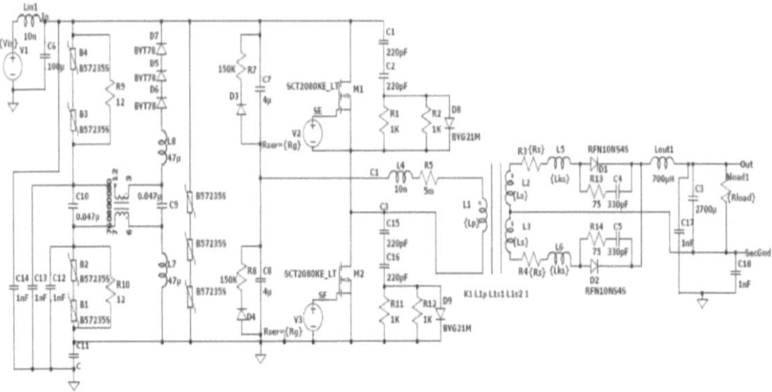

Figure 63 Basic schematic of prototype SiC DC-DC converter

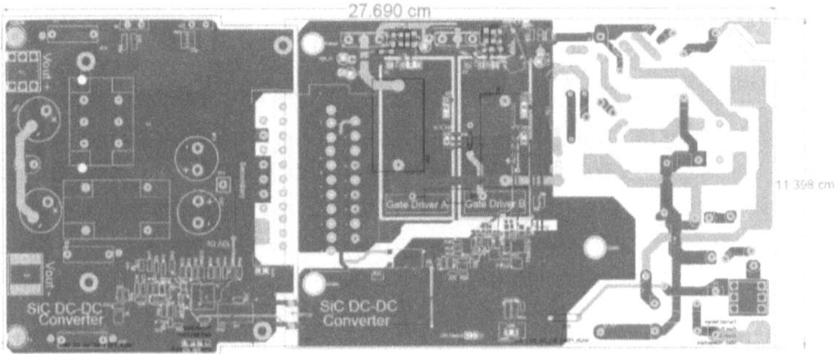

Figure 64 PCB layout for the prototype converter

73

Figure 65 Gate driver modules testing in prototype DC-DC converter

Because of the challenge to test all the designed drivers in the prototype, PCBs are designed in a way to achieve the plug and play mode for driver modules. These gate drivers are placed with a minimum distance to the MOSFETs to minimize the gate loop inductance. Furthermore, few test points are added just next to the MOSFETs and gate drivers. These placements for modules and test points contribute to only 5nH of loop inductance in the prototype thus providing a clean and error free measurements.

7 Experimental validation of gate drivers in prototype converter

At first the constructed SiC based half-bridge prototype is tested for basic operation. The duty cycle and output load are adjusted to check overall efficiency of the converter on different stages. The low circuitry and gate drive circuits are tested to confirm the gate to source voltages present for the SiC MOSFETS. The two complementary logic signals are supplied externally to the gate drive modules with the high speed 16-bit microcontroller from DSPIC 33FJ64MC802.

7.1 Basic operation test

Basic gate driver functionality is confirmed in the Figure 66. To operate prototype DC-DC converter's gate drivers are supplied with DC voltage 12V externally. The gate-source-voltage V_{GS} is measured with the SiC MOSFET gate and source pins directly. The +15V/-5V at V_{GS} is confirmed and the round curves on the waveform show the charging the C_{GD} and C_{GS} capacitors within the MOSFETs. After successful gate drive and low circuitry test basic operation of designed converter is tested with output load from 1-9A on different frequency levels.

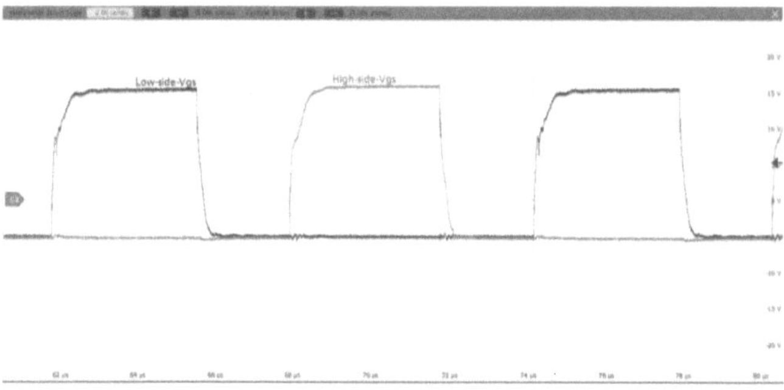

Figure 66 Gate drive functionality test in prototype converter

The initial testing started with 10V as input voltage and slowly ramping up the input and output power levels. The power level and duty cycle are slowly adjusted to achieve fail safe operation in the converter. For the initial testing a duty cycle is set to D 0.47 with suitable dead time between both high and low side switching transients. The Figure 67 shows the captured waveforms at 100V in input, duty cycle 0.47, 80 kHz and 100W output power level. Input voltage of power converter is adjusted by 10kW power supply and the load at output is adjusted in steps of approximately ~1kW. To minimize the differences in the testing platform, same setup and types of probes are used in all testing.

Figure 67 Primary side waveforms of prototype SiC DC-DC converter

In initial testing overall efficiency is measured around 89% due to some overheating components present in the input filter at primary side. Moreover, the snubber circuit at secondary side needs adjustments. After adding the 200pF capacitor, the oscillation frequency is changed from 3.87MHz to 2.42MHz as Figure 69(a). At CP 330pF the oscillation frequency is reduced by factor of two which is around 2.17MHz. Afterwards, the value of the R_{sn} =0.76 Ω is calculated from the section 6.2.5. Overall efficiency of the converter is increased by 3% after adding primary and secondary snubbers. The Figure 69(b) shows the voltage at diodes after adding the snubber circuit.

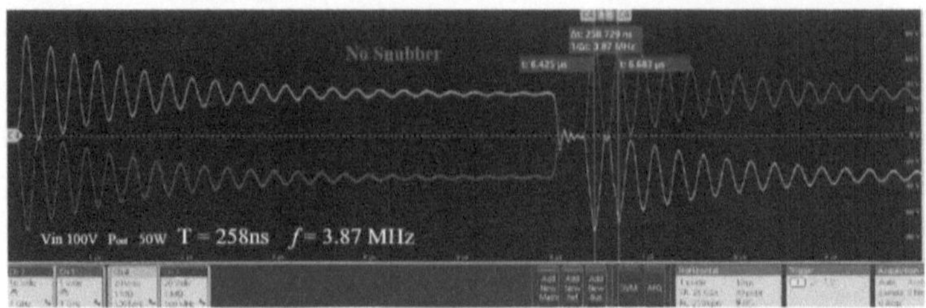

Figure 68 The voltages across rectifier diodes without snubber circuit

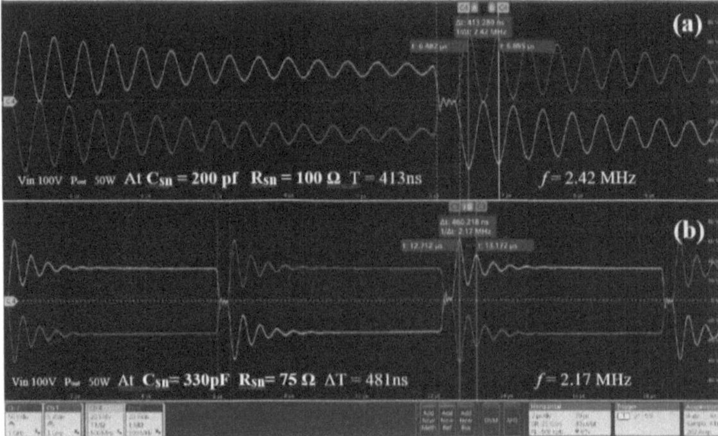

Figure 69 The voltages across rectifier diodes (a) at 200Pf & 100Ω (b) at 330Pf & 75Ω

The temperature of individual components is monitored in early testing of prototype to identify the potential issues. The three diodes BYT78 present in the input filter from Vishay for surge protection shows high heat dissipation. In the final testing these diodes are shorted to achieve the best efficiency of the with the SiC MOSFETs and designed gate drivers. The Figure 70 shows the heating diodes present in the schematic while Figure 72 shows the heating image captured from the camera with the diodes.

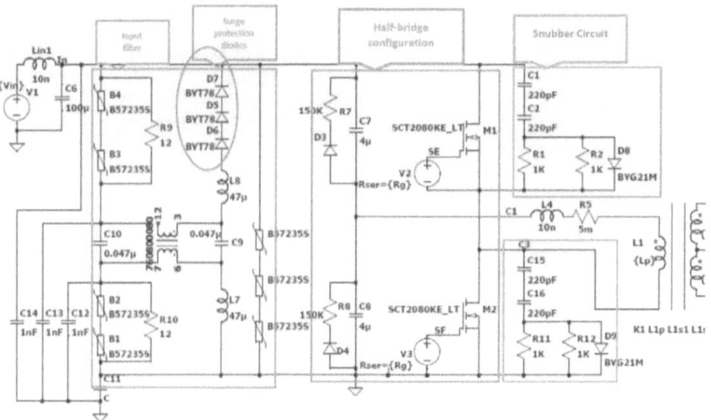

Figure 70 Primary side surge protection diodes

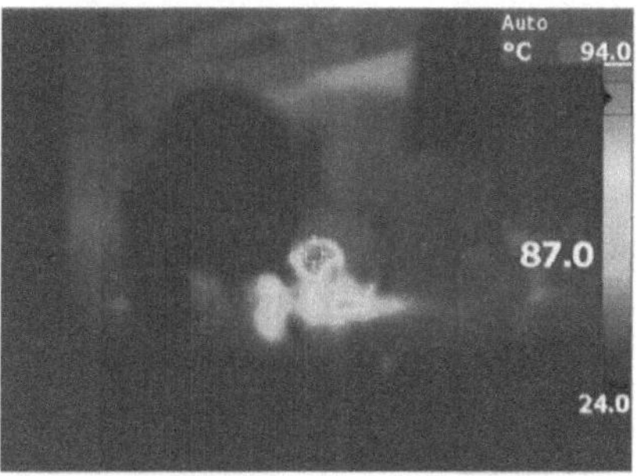

Figure 71 Heat dissipation present surge protection diodes (BYT78)

After successful addition of snubber circuit and shorting the surge diodes the overall efficiency becomes higher reaching around 95%. The green and red waveform shows the voltage at diodes and blue waveform represents the average output voltage after output filter. The equation below is verified with the set duty cycle and output voltage. This results into rated output voltage 110V. Slight difference in the peaks of diode voltages is due to using the differential probs with different bandwidth. The output voltage around 110V can be also verified in the Figure 72.

$$V_{out} = \frac{P_1}{P_2} D \times V_{in}$$

(1.41)

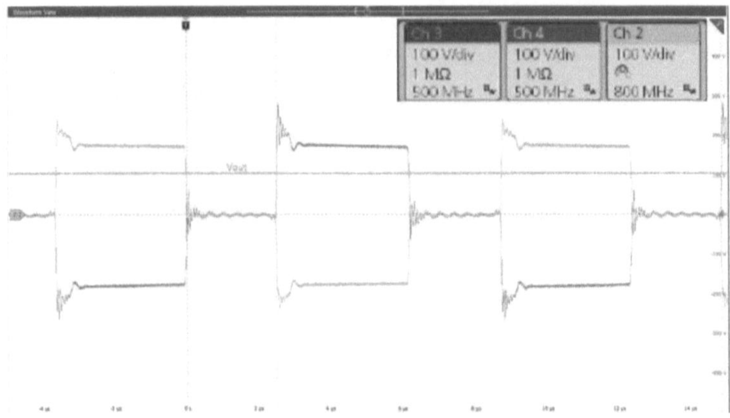

Figure 72 Waveforms for Secondary side of prototype

7.2 Gate drivers test in designed converter

7.2.1 Miller effect avoidance

The false turn-on effect happens due to high dV/dt observed in the SiC devices, which can exceed 50kV/μs. The Miller clamp feature in the gate driver modules (b),(c) provide a lower impedance path without compromising the slew rates for turn-off. Often false turn-on problem can arise if the same value of gate resistors is used for both turn-on and turn-off [13], thus resistance of the gate drive turning the device on should be much higher than that the one holding the opposing transistor off. A typical ratio might be anywhere from 4:1 or 2:1.

The gate drivers used in our research mainly helps in the Miller avoidance as depicted in the Figure 73. Subject to the condition that there is no excessive overshoot on the drive and the Miller effect mentioned in the section 2.6.6 does not contribute.

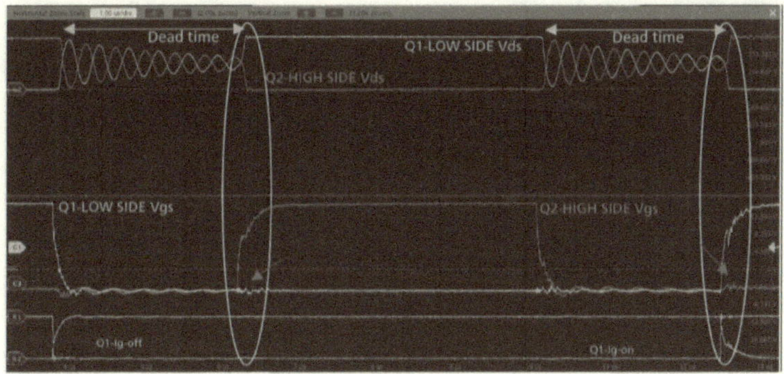

Figure 73 Switching waveforms for designed SiC half-bridge configuration with gate driver (c) ADuM4135

It is just as important that the device is properly held off. There is no need for a Miller clamp as long as there is a sufficient negative gate drive margin to keep the devices held off, which ties back to the requirement for the negative drive. As the driver (a) doesn't feature the Miller clamp function but V_{GS} bump is around ΔV_{GS} =2V.

7.3 Gate drive modules tests

Constructed DC-DC converter is tested at different power levels with all the designed gate driver circuits. While comparing the switching transients, the following parameters are kept equal: gate voltage (+ 15 V and - 5 V), drain voltage (500 V) and drain current (7.5 A) at 80kHz with a duty cycle of 0.30. The used transformer in the circuit is rated for 100 kHz therefore, the measurements are carried out at two frequency levels i.e. 80-100 kHz. As a hard-switched converter, the duty cycle is adjusted to D 0.30 to achieve set level voltage. The results present out a very high dead time due to limitation of transformer and application requirement. The figures below present the V_{GS} and V_{DS} and I_G waveforms with the gate driver Ucc27531 at primary side at 750W output. The gate current of individual module is also measured in the prototype to evaluate the drive strength of designed modules in designed prototype.

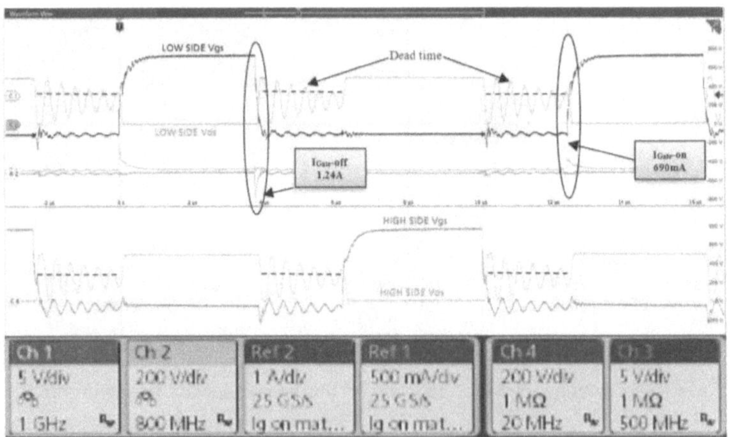

Figure 74 Primary side switching waveforms with gate driver A (Ucc27531)

The main aspect in half bridge topologies with SiC MOSFETs is to over-come the risk of parasitic turn-on of the upper MOSFET during the switching of lower device, and vice versa. The Ucc27531 without having a Miller clamp feature helps in keeping the gate voltage at low due to asymmetric gate to source voltages such as +15/-5V. Moreover, the gate resistor ratio selection from previous section 5.5 helps in providing the balance with fast transitions and lower EMI.

The Figure 75 shows the waveforms of the primary side V_{GS}, V_{DS}, and gate current (I_G) within a same waveform in a single output gate driver IED020I12-F2. Results conducted from previous measurements clear that the rising and falling slew rates for IED020I12-F2 are comparatively slow, therefore, a smaller gate resistor Rg 10Ω is selected. Compared to the other gate driver the gate driver (b) leads in generating less EMI but provides slightly bigger slew rates and delay from input to output. Which will ultimately produce high switching losses and high switching stress on the DUT.

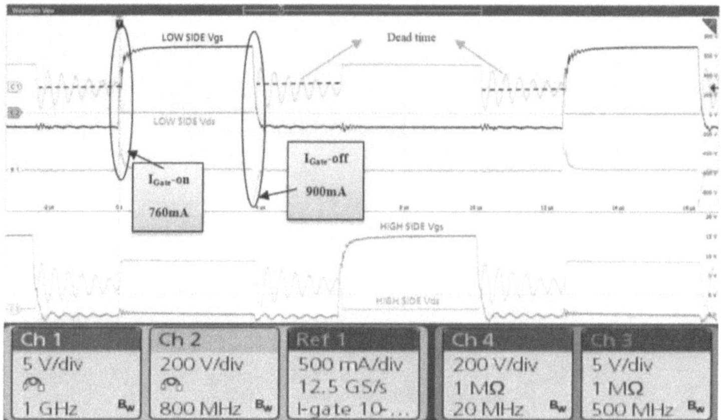

Figure 75 Primary side switching waveforms with gate driver B (IED020I12-F2)

The switching waveforms at primary side with the driver module (c) show the faster ramps and higher drawn gate current for switching transitions. While in the advanced gate drivers like ADUM4135 and 1ED020I12-F2 Miller clamp circuit is separately designed within the chip. In this circuit, the current charging the C_{GD} Miller capacitance during high dV/dt is shunted by the pulldown stage of the driver. A separate pulldown MOSFET is utilized within the chip to achieve clamp functionality. A voltages spike can result in the V_{GS} of the MOSFET when the pulldown impedance is not low enough. During turn-off, the gate voltage is monitored, and the clamp output is activated when the gate voltage goes above 2 V.

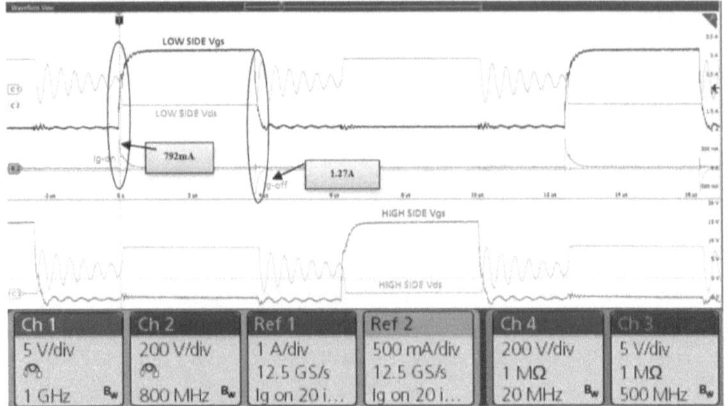

Figure 76 Primary side switching waveforms with gate driver C (ADuM4135)

7.3.1   The gate current measurements in prototype

The behaviour of gate current in prototype is also measured to verify the results achieved from DPT measurements. The turn-on gate resistance is 20Ω for Ucc27531 and ADuM4135. 1ED020I12-F2 is tested on gate resistance 10Ω to minimize rise slew rates and a big delay is observed in previous studies, thus providing more current in turn-on transition. The turn-off resistance is 10Ω for all gate drivers. The highest peak sink and source current are captured with the gate driver ADuM4135.

Figure 77 The gate current Ig-on(p-p) seen in DC-DC converter with driver modules

Figure 78 The gate current Ig-off (p-p) seen in DC-DC converter with driver modules

Most importantly, the gate driver C from Analog devices (ADuM4135) showed the best current capability by providing 1.27A peak to peak gate current in turn-off period. The gate driver (a) is ranked the second highest overall by providing Ig-p-p of 1.205A.

7.4 Switching loss comparison in SiC DC-DC converter

Furthermore, power losses and thermal behaviour of the power module are observed. Number of tests for various frequencies and configurations of load are carried out in addition to different gate driver performances and switching losses with different gate drivers. SiC MOSFETs show losses during switching transients due to voltage and current overlapping and the losses due to Rds(on) in on state. The switching losses comparison ranging from 80-120kHz is shown in the figure below. The switching losses increase as frequency increases while conduction losses remain same as due to only duty cycle dependency. The prototype offers very high dead time losses due duty cycle and output relationship.

7.4.1 Switching loss estimation

The switching energies and losses in the MOSFET on the primary side are calculated from the section 2.8. Where the conduction losses are:

$$P_{cond} = Id_{rms}^2 \cdot Rds^{ON} \tag{1.42}$$

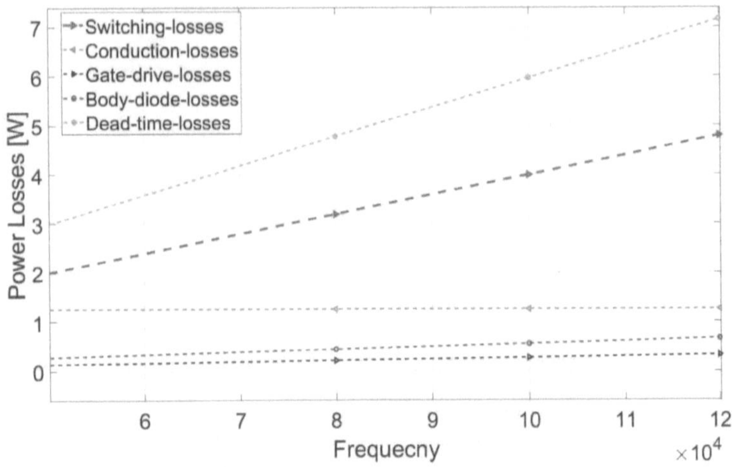

Figure 79 Power losses estimation in prototype

RMS current through MOSFET is:

$$Id_{rms} = \sqrt{D} \times \left(\frac{Id_{min}+Id_{max}}{2}\right) \qquad (1.43)$$

$$Id_{rms} = \sqrt{0.3} \times \left(\frac{4+5.5}{2}\right)$$

The conduction losses per switch are:

$$P_{cond} = 0.5408W$$

The turn-on and off losses in MOSFET are:

$$P_{SW} = \left(\frac{V_{IN}.Id_{min}\ t_r}{6T} + \frac{V_{IN}.Id_{max}\ t_f}{6T}\right) \qquad (1.44)$$

$$P_{SW} = \left(\frac{280\times4\times51\times10^{-9}}{6\times12.55\times10^{-6}} + \frac{399\times5.5\times51\times10^{-9}}{6\times12.55\times10^{-6}}\right) = 2.24W$$

The gate charge losses can be from the equation (1.09):

$$P_G = 2(180 \times 10^{-9}) \times 15 \times 80 \times 10^3 = 432mW$$

Dead time losses can be calculated from the equation (1.11):

$$P_D = 4.6 \times 2.7 \times (2.4 + 2.4) \times 10^{-6} \times 80 \times 10^3 = 4.7W$$

The switching losses are minimized by using the suitable possible ON and OFF drive. The choice of gate resistance is made from the previous section 5.5 and chosen to be Rg-off 10Ω and Rg-on 20 Ω. The figures below present the primary side voltages and MOSFET power losses on the low side with the different gate drivers at 750W output at 80kHz. The losses increase with increasing the switching frequency. The Figure 80-82 demonstrates the switching losses in low side of the prototype with different driver modules.

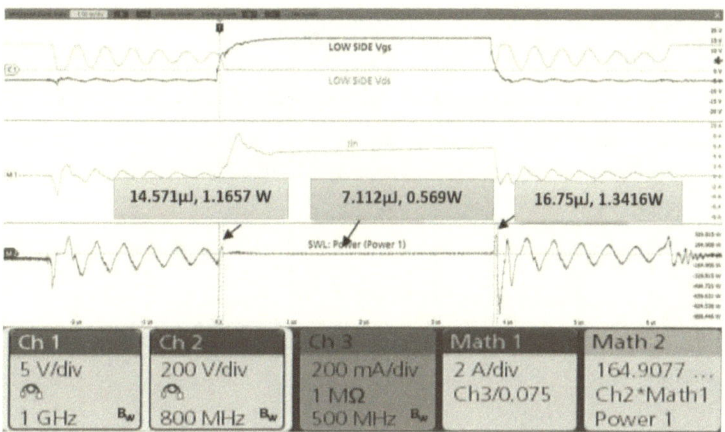

Figure 80 Switching losses in SiC MOSFET with gate driver A (Ucc27531)

Valley ringing can be seen due to high dead time present during switching intervals. However, maintaining the dead time larger than the propagation delay results in reduced system efficiency due to current flowing back through the body diode during the dead time. Thus, in dead time losses significantly increases as the voltage drops across the diode becomes large [40].

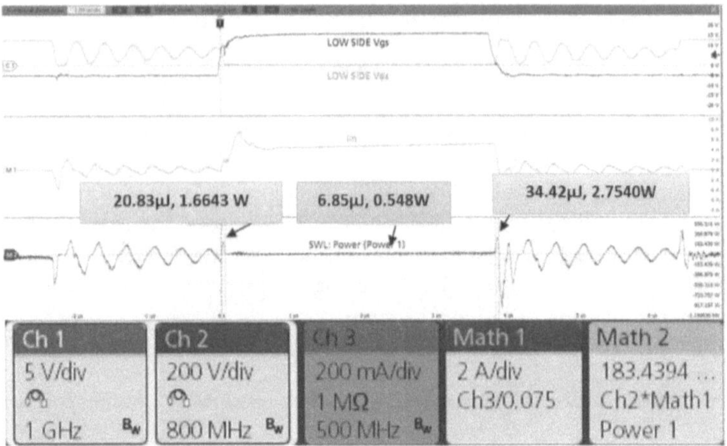

Figure 81 Switching losses in SiC MOSFET with gate driver IED020I12-F2

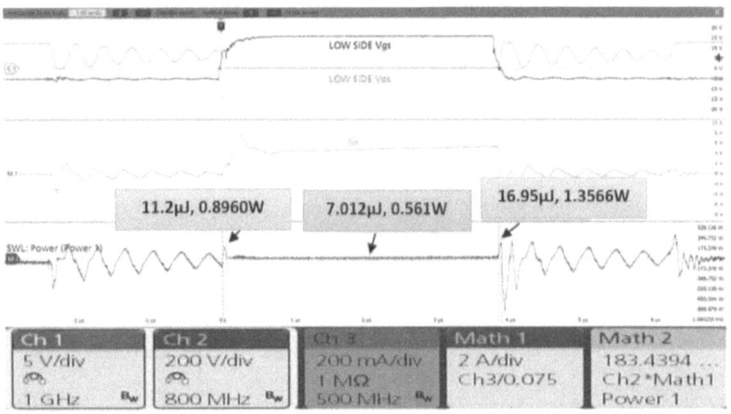

Figure 82 Switching losses in SiC MOSFET with gate driver C (ADuM4135)

But as found out in the previous studies (5.3) the SiC MOSFETs offers the reverse recovery of used SiC MOSFETs is almost 10ns thus, turn-on losses are notably lowered due to the Qrr of the SiC diode. For the gate driver module (c) the switching losses for turn-off and turn-on are about 0.8960W and 1.3566W respectively, whereas the conduction losses around 0.596W are much alike in all gate driver measurements due to duty cycle and R$_{DS}$(on) temperature dependency. The Table 23 provides the details switching losses calculation with different driver modules.

Table 23 Switching Energy losses comparison with different gate drivers

Gate Driver	E-on (W)	E-off (W)	E-Cond
Ucc27531	14.571μJ	16.75μJ	7.112μJ
1ED020I12-F2	20.83μJ	34.42μJ	6.85μJ
ADuM4135	11.2μJ	16.95μJ	7.012μJ

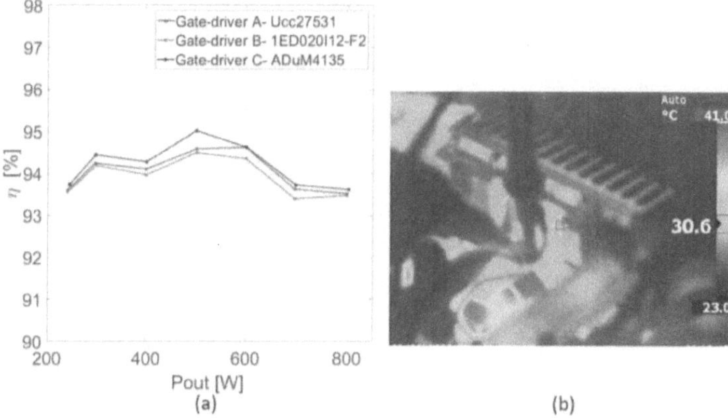

Figure 83 (a) efficiency comparison with driver modules (b) SiC MOSFETs temperature with heat sinks.

The slightly better efficiency can be observed with the gate driver ADuM4135, where Ucc27531 and 1ED020I12-F2 follow the same trend, respectively. But only very slight change in overall efficiency can be observed in the gate driver (a) and (c) due to providing very close characteristics. Moreover, the temperature of the gate driver modules ranges around 50 C° whereas SiC MOSFETs are observed around 25-30C°. The Figure 83 illustrate the efficiency comparison of different gate drivers. The measurements are conducted as close as possible to the full load.

8 Discussion

Wide band gap in SiC MOSFETs helps in reaching the higher voltage levels and providing the lower switching losses. Different driver solutions for SiC MOSFETs are implemented. The driver modules and prototype design are further optimized by parasitic minimization placement. The SiC MOSFET from ROHM provides a balance in slew rates and EMI generated in switching periods. It can be seen in Figure 84, the gate driver modules Ucc27531 and ADum4135 parameters remain fairly similar and capable to provide the rise and fall times needed to charge and discharge the capacitance. The designed driver modules (a) and (c) are capable of providing the rise/fall times less than 20ns and 60ns respectively. The driver modules use the optocoupler technology which provides a propagation delay around 50ns.

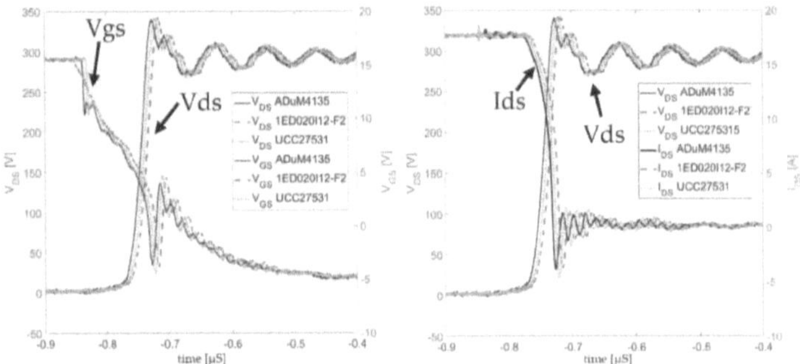

Figure 84 The gate driver results in DPT

Both driver modules help in reducing the power losses to minimum. The gate driver Ucc27531 shows more ringing in the gate voltage and does not help in protection features. ADuM4135 helps in reducing the overshoots, EMI, and specially helps in turn-off transition by providing a lower impedance path to sink the current. All the gate deriver modules provide consistency in the temperature performance.

Whereas the driver (b) uses a galvanic isolation technique thus results into propagation delay around 160ns and slower rise and fall time due to low sink and source capability. The gate driver 1ED020I12 shows minimum sink and source current and slower slew rates but provides less EMI in the transitions.

The gate driver modules test shows the same performance in the prototype converter. The sink and source current used by the MOSFETs in the converter are much smaller than the double pulse test. This is due to lesser dI/dt requirements in the converter. All the gate drivers perform alike in efficiency comparison. Overall efficiency improvement is seen in driver (a) and (c). The prototype converter shows the very good temperature performance overall.

Furthermore, the overall efficiency is also compared with conventional DC-DC converter. The conventional converter is configured with Si MOSFETs and conventional gate drivers. The Si MOSFETs are rated for 580 V_{DS} and I_{DS} 4A. Considering all the parameters the discrete comparison is achieved shown in the figure below. The overall waveforms are compared with conventional DC-DC converter. Both converters are tested on the same frequency, duty cycle and using the same transformer to compare design in the depth. The conventional converter is configured with gate drivers LM5112 from Texas instruments. The gate driver is built with 0 to 12V gate to source voltage functionality without negative voltage option. The waveforms are observed at primary LOW side of SiC based dc to dc converter.

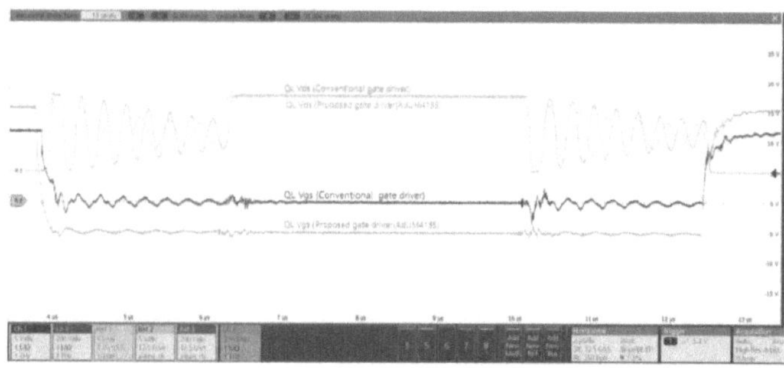

Figure 85 Comparison results with conventional vs proposed design

Here, the fast turn off and turn on can be observed with minimum slew rates. The reduced valley ringing shows less losses generated in dead time due to an optimal design. The dead time losses are reduced because of fast diode reverse recovery of SiC devices. Due to low rated current values and high thermal dissipation the highest output power shows the decrement in efficiency.

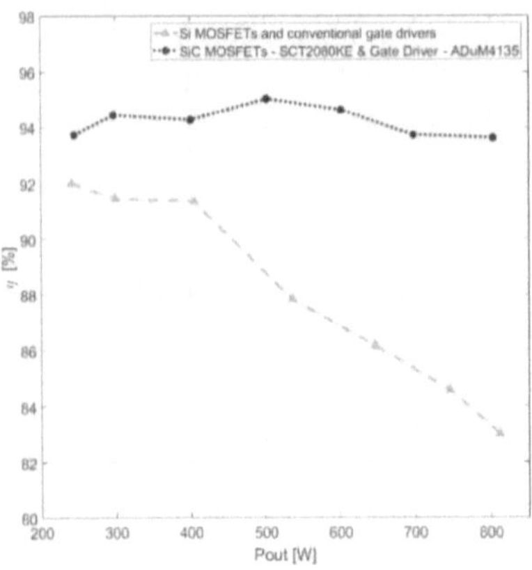

Figure 86 Efficiency comparison in Si vs SiC DC-DC converter

The actual efficiency of prototype converter can be better if the input filter and other components attached before the half bridge configuration are not present. The overall filter losses are about 3%, giving the peak efficiency to 94.5%. The overall efficiency comparison graph can be seen in the Figure 86. This results in optimally less EMI generated in the switching transitions and by adopting to negative voltage the probability of miller effect is almost none. While in conventional gate driver (LM5112) gate voltage miller spike is reaching to almost 4V, which is close to threshold of gate to source voltage of the switch. Due to low rated current values and high thermal dissipation, the highest output power shows the decrement in efficiency.

8.1 Social aspects

The conventional switch mode power supplies work at lower frequencies due to limitations in switches and transformers. The SiC MOSFETs with designed gate drivers can operate at higher frequencies thus results in smaller peripherals like transformer and filter design. Due to working at such higher frequencies it is important to fulfil the EMI requirements which could be very challenging. The overall design can operate at elevated temperatures by still maintaining the lower energy loss in the system. All these advancements in converter effectively help in reducing the switching losses thus providing a cleaner and safer environment. The designed gate drivers offer best solution to the industry, which is focusing to acquire energy efficient systems.

8.2 Ethical aspect

Highest ethical and professional conduct was followed during the research. The components and materials used in design are RoHS Compliance, lacking the traces of any heavy metal including Lead (Pb), Cadmium (Cd), and Mercury (Hg). Proper public safety protections were abided in the work.

9 Conclusions

Suitable gate drive circuit for SiC-MOSFETs is designed and further analysed, which is capable to deliver adequate output voltages wing at switching frequencies up to fsw =500kHz. The gate driver provides maximum current to operate the device with enough protection features. The available SiC-MOSFETs are compared by using simulation and real-time based measurements. Here, three different drivers A (Ucc27531), B (1ED020I12-F2), and C (ADuM4135) are compared to find out the best gate driver IC available in the market. The effects related to the switching transients including switching delay, noise generation, voltage overshoot, and oscillations are investigated. The gate driver ADuM4135 performs much quicker and shows the minimum over and undershoots in the switching transients. The driver module uses iCoupler technology and provides a minimum input to output delay.

The driver can provide the peak gate current required for a fast and safer transition. The driver also shows the best performance in the short circuit protection by providing a 300ns masking time to provide a softer turn-off. Furthermore, the gate driver helps in Miller effect avoidance by providing a lower impedance path in miller plateau region. This helps to minimize the gate voltage spike observed in the testing and reducing the false turn-on problems.

The simulation verification of DPT tester is carried out and optimized to verify the work. It is observed in the simulation that the drain to source (V_{DS}) rather shows a different delay time as shown in the measurement and data sheets due to problems in the simulation models provided by the manufacturers. Finally, the complete verification of a DC-DC converter is integrated with ADuM4135 gate driver and further tested. Here, it is interesting to observe that the overall efficiency is somewhat different compared to other driver modules due to other heavy losses present in the design. In combination with the low module inductance design, higher peak current and fast ramp up provided by the driver, the overall converter efficiency including passive components is increased to 94.5%, better to the previously designs of the conventional converters. The designed gate drivers and prototype converter provides all the attractive features and can be widely implemented in industrial applications for energy efficient systems.

9.1   Future work

Future enhancements can include testing the gate drivers on higher frequency applications, at high temperatures for extended period to analyse the robustness of the design. Electromagnetic interference comparison with conventional gate drivers should be performed to check if it is being affected. Although, ADuM4135 supplies high current but further improvements can be made via using higher power supplies and gate- to-source voltage of +18/-5V which would result in lower $R_{DS}(on)$. Further evaluation is possible by using the gate driver modules in advanced GaN based bi-directional converters. Furthermore, the gate drivers could be integrated with the same package as SiC power MOSFET device. Integration in the same package/chip would result unmatched performance potential.

References

[1] "Silicon Carbide Market," 02 2020. [Online]. Available: https://www.eetimes.eu/sic-mosfets-for-the-next-generation-of-evs/?fbclid=IwAR1YbvNdpUE27g9rhHk1Hcp6XxuEsy-SKrjKKg2ZxllhSkUMH3yUdMG_BDE. [Accessed 05 01 2021].

[2] R. Rupp, "CoolSiC™ and major trends in SiC power device development," in *European Solid-State Device Research Conference (ESSDERC)*, Leuven, 2017.

[3] T. Sakaguchi, M. Aketa, T. Nakamura, M. Nakanishi and M. Rahimo, "Characterization of 3.3 kV and 6.5 kV SiC MOSFETs," in *PCIM Europe*, Nuremberg, Germany, 2017.

[4] B. S. Naik, S. Shan, L. Umanand and B. S. Reddy,, "A Novel Wide Duty Cycle Range Wide Band High Frequency Isolated Gate Driver for Power Converters," in *IEEE Transactions on Industry Applications*, 2018.

[5] J. Zhao, J. Zhang, H. Wu and Y. Zhang, "Design of Gate Driver and Power Device Evaluation Platform for SiC MOSFETS," in *WIPDA ASIA*, Xian, 2018.

[6] A. Gendron-Hansen, "High-Performance 700 V SiC MOSFETs for the Industrial Market," in *WiPDA*, NC,USA, 2019.

[7] M. Imaizumi , "Remarkable advances in SiC power device technology for ultra high power systems," in *IEEE International Electron Devices Meeting*, Washington, DC, 2013.

[8] A. Elasser and T. P. Chow, "Silicon carbide benefits and advantages for power electronics circuits and systems," in *IEEE*, 2002.

[9] J. Dodge, "Power MOSFET Tutorial," B - Advanced Power Technology, 2006.

[10] N. K. Pilli and S. K. Singh, "Influence of peak gate current and rate of rise of gate current on switching behaviour of SiC MOSFET,," in *IEEE Transportation Electrification Conference (ITEC-India*, Pune, 2017.

[11] ROHM, "Datasheet SCT2080KE SiC MOSFET," 2018.

[12] V. Barkhordarian, Power MOSFET Basics, El Segundo, Ca: International Rectifier.

[13] C. Stewart, A. Escobar-Mejía and J. C. Balda, "Guidelines for developing power stage layouts using normally-off SiC JFETs based on parasitic analysis," in *EEE Energy Conversion Congress and Exposition*, Denver, 2013.

[14] Wang, Fei, Characterization of Wide Bandgap Power Semiconductor Devices, Institution of Engineering & Technology, 2018.

[15] J. Balcells and P. Bogónez-Franco, "Effect of driver to gate coupling circuits on EMI produced by SiC MOSFETS," in *International Symposium on Electromagnetic Compatibility*, Brugge, 2013.

[16] Y. Gao, A. Huang, S. Krishnaswami, J. Richmond and A. Agarwal, "Comparison of Static and Switching Characteristics of 1200 V 4H-SiC," *IEEE Transactions on Industry Applications*, pp. 887-893, 2008.

[17] Y. Zhu, H. Li, C. Luo, Y. liu, C. Wan and J. Ma, Influence of Paralleled SiC MOSFET on Turn-off Gate Voltage Oscillation, Detroit: ECCE, 2020.

[18] C. DiMarino, J. Wang, R. Burgos and D. Boroyevich, " high-power-density, high-speed gate driver for a 10 kV SiC MOSFET module," in *2017 IEEE Electric Ship Technologies Symposium (ESTS)*, Arlington, VA , 2017.

[19] T. Ishigaki, "Analysis of Degradation Phenomena in Bipolar Degradation Screening Process for SiC-MOSFETs," in *ISPSD*, Shanghai, China, 2019.

[20] N. X. Huang, M. S. Shiau, H. Wu and D. G. Liu, "Improving stress effect and low noise design of the integrated on gate driver,," in *MIXDES 2010*, Warsaw, 2010.

[21] H. Kim, S. Acharya, A. Anurag, B. Kim and S. Bhattacharya, "Effect of Inverter Output dv/dt with Respect to Gate Resistance and Loss Comparison with dv/dt Filters for SiC MOSFET based High Speed Machine Drive Applications," in *ECCE*, Baltimore, 2019 .

[22] STMicroelectronics, "Applciartion note-AN5355-Mitigation technique of the SiC MOSFET gate voltage glitches with Miller clamp," ST, 2019.

[23] S. R. Jang, S. H. Ahn, H. J. Ryoo and G. H. Rim, "A comparative study of the gate driver circuits for series stacking of semiconductor switches," in *EEE International Power Modulator and High Voltage Conference, Atlanta*, Atlanta, 2010.

[24] STMicroelectronics NV, "Mitigation technique of the SiC MOSFET gate voltage glitches with Miller clamp," Application note , 2019.

[25] Rohm,Application note, " Gate-source voltage behaviour in a bridge," 2016.

[26] L. Guillaume, D. B. Viet, F. J. Paul, B. Jean and L. Yves, "New soft switching ZVS and ZCS half-bridge inductive DC-DC converters for fuel cell applications," in *IEEE International Power Electronics Congress*, Puebla, 2006.

[27] R. W. Erickson and D. Maksimovic, Fundamentals of Power Electronics, Secaucus, NJ: Kluwer Academic Publishers, 2000.

[28] Ferreira, X. Gong and J. A., "Comparison and Reduction of Conducted EMI in SiC JFET and Si IGBT-Based Motor Drives," *IEEE Transactions on Power Electronics*, pp. 1757-1767, April 2014.

[29] J. V. P. Sekhar, R. Maheshwari and H. Li, "An energy recovery based low loss resonant gate driver circuit for Silicon Carbide MOSFETs," in *IEEE International Conference on Power Electronics, Drives and Energy Systems (PEDES)*, Trivandrum, 2016.

[30] K. V. Nguyen and N. Nguyen-Quang, "Novel Tiny 1.2kV SiC MOSFET Gate Driver," in *ide Bandgap Power Devices and Applications in Asia (WiPDA Asia)*, Xi'an, China, 2018.

[31] Texas Instruments, "Data sheet- Ucc27531," Dallas, Texas , 2018.

[32] Infineon Technologies AG, "DATA SHEET-EiceDRIVE-1ED020I12-F2," 2017.

[33] Analog devices, "DATA SHEET- ADuM4135- High Voltage Isolated gate driver".

[34] Murata power solution, "Data sheet- MGJ2 Series-5.2kVDC Isolated 2W Gate Drive DC-DC Converters".

[35] D. Kim, D. Joo, B. Lee and J. Kim, "Design and analysis of GaN FET-based resonant dc-dc converter," in *ECCE Asia (ICPE-ECCE Asia)*, Seoul, 2015.

[36] Ferreira, X. Gong and J., "Comparison and Reduction of Conducted EMI," in *IEEE Transactions on*, 2013.

[37] Z. Chao, T. Xiao-Yan, Z. Yu-Ming, W. Wen and Z. Yi-Men, "a DC-DC boost converter based on SiC MOSFET and SiC SBD," in *IEEE International Conference of Electron Devices and Solid-State Circuits*, Tianjin, 2011.

[38] B. J. Baliga, Fundamentals of Power Semiconductor Devices, BOSTON USA: Springer US, 2008.

[39] A. M. S. Al-bayati, S. S. Alharbi, S. S. Alharbi and M. Matin, "A comparative design and performance study of a non-isolated DC-DC buck converter based on Si-MOSFET/Si-Diode, SiC-JFET/SiC-schottky diode, and GaN-transistor/SiC-Schottky diode power devices," in *North American Power Symposium (NAPS)*,, Morgantown WV, 2017.

[40] S. S. Alharbi, S. S. Alharbi and M. Matin, "An Improved Interleaved DC-DC SEPIC Converter Based on SiC-Cascade Power Devices for Renewable Energy Applications," in *IEEE International Conference on Electro/Information Technology (EIT)*, Rochester, MI, 2018.

[41] H. Kato , "Comparative analysis of full bridge and half bridge current resonant DC-DC converter,," in *IEEE 33rd International Telecommunications Energy Conference (INTELEC)*, Amsterdam, 2011.

[42] L. Balogh, "L. Balogh," kassel university press Gmbh, Kassel, 2007.

[43] J. J. Graczkowski, K. L. Neff and X. Kou, "A Low-Cost Gate Driver Design Using Bootstrap Capacitors for Multilevel MOSFET Inverters," in *CES/IEEE 5th International Power Electronics and Motion Control Conference*, Shanghai, 2006.

The MATLAB scripts below use to characterize gate drivers in double pulse
test as seen in   Figure 44 . While the second script is used to calculate the
switching losses estimation as shown in the section 7.4.1.

```matlab
%for k=1:length(file)
fnames={'Adum r20 r20 ig on_ALL.csv','inf r20 _ALL.csv',...
        'ti r 20 20 i g on _ALL.csv',}
figure(1)
clf
for FileK=1:length(fnames)
res=dlmread(fnames{FileK},',',11,0);
t=res(:,1).*1e6;
Vgs=res(:,2);
Vds=res(:,3);
Ids=res(:,6);
VgsMax=max(Vgs);
VgsMin=min(Vgs);
dVgs=VgsMax-VgsMin;
Vtoff=VgsMax-dVgs*0.1;
Vton=VgsMin+dVgs*0.1;
k=find(Vgs<Vtoff);
toff=t(k(1));
k=find(Vgs>Vton & t>.5);
toff=t(k(1));
subplot(321)
plot(t-toff,Vgs)
legend('AD-R5','AD-R30')
axis([-0.01 0.3 -10 20])
grid on
lgd = legend('UCC27531','1ED020I12-F2','ADuM4135');
ylabel('V_G_S [V]')
hold on
subplot(323)
plot(t-toff,Vds)
lgd = legend('UCC27531','1ED020I12-F2','ADuM4135');
title('Vds Comp Turn OFF')
axis([-0.01 0.3 0 400])
grid on
ylabel('V_D_S [V]')
hold on
subplot(325)
plot(t-toff,Ids)
lgd = legend('UCC27531','1ED020I12-F2','ADuM4135');
axis([-0.01 0.3 -10 20])
title('Id Comp  Turn OFF')
grid on
xlabel('time [\mus]')
ylabel('I_D_S [A]')
hold on
subplot(322)
```

```matlab
plot(t-toff,Vgs)
lgd = legend('UCC27531','1ED020I12-F2','ADuM4135');
axis([-0.01 0.63 -10 20])
grid on
title('Vgs Comp  Turn ON')
hold on
subplot(324)
plot(t-toff, Vds)
lgd = legend('UCC27531','1ED020I12-F2','ADuM4135');
title('Vds Comp  Turn ON')
axis([-0.01 0.63 0 400])
grid on
hold on
subplot(326)
plot(t-toff,Ids)
lgd = legend('UCC27531','1ED020I12-F2','ADuM4135');
FontSize = 30;
title('Id Comp  Turn ON')
axis([-0.01 0.63 -2 20])
grid on
xlabel('time [\mus]')
hold on
end
Vgs on
figure(1)
for FileK=1:length(fnames)
res=dlmread(fnames{FileK},',',11,0);
t=res(:,1).*1e6;
Vgs=res(:,2);
Vds=res(:,3);
Ids=res(:,6);
VgsMax=max(Vgs);
VgsMin=min(Vgs);
dVgs=VgsMax-VgsMin;
Vtoff=VgsMax-dVgs*0.1;
Vton=VgsMin+dVgs*0.1;
k=find(Vgs<Vtoff);
toff=t(k(1));
k=find(Vgs>Vton & t>.5);
toff=t(k(1));
plot(t-toff,Vgs)
lgd = legend('UCC27531','1ED020I12-F2','ADuM4135');
axis([-0.01 0.3 -10 20])
grid on
ylabel('V_G_S [V]')
xlabel('time [s]')
set(gca,'fontsize',18)
set(gcf,'color','white')
hold on
axis square
end
```

Switching losses calculations:

```matlab
Calculations for Switching losses
%Switching losses
Rdson=80*10^(-3);
```

```matlab
D 0.5;
%Gate driver ADum4135 parameters
Idmina=1.6;
Idmaxa=4.7;
Vinra=399;
Vinfa=280;
tra=51*10^(-9);
tfa=60*10^(-9);
fa=80*10^(3);
Ta=1/fa;
% Gate driver UCC27531 parameters
Idminu=1.6;
Idmaxu=4.8;
Vinru=406;
Vinfu=273;
tru=46.7*10^(-9);
tfu=78*10^(-9);
fu=80*10^(3);
Tu=1/fu;
% Gate driver INF-F2 parameters
Idminf=2;
Idmaxf=4.8;
Vinrf=405;
Vinff=395;
trf=102*10^(-9);
tff=79*10^(-9);
ff=80*10^(3);
Tf=1/ff;
% Conduction Losses
Idrms=(sqrt(D))*((Idmina+Idmaxa)/2);
Pcond= (Idrms)^(2)*Rdson
Pconta= 2*Pcond
% ON switching losses
Ta=1/fa;
Tf=1/ff;
Tu=1/fu;
Pswra= ((Vinra*Idmina*tra)/(6*Ta))
Pswfu= ((Vinfu*Idminu*tfu)/(6*Tu))
Pswff= ((Vinff*Idminf*tff)/(6*Tf))
% OFF switching losses
Pswra= ((Vinra*Idmina*tra)/(6*Ta))
Pswru= ((Vinru*Idminu*tru)/(6*Tu))
Pswrf= ((Vinrf*Idmaxf*trf)/(6*Tf))
Tri=31*10^(-9)    % taken the reverse recovery time
Tfi=22*10^(-9)
trr1=31*10^(-9)
Qrr=44*10^(-9)
Rds = 80*10^(-3)
fsw=80*10^(3)
Udr=15
Uplt=9.7
Rgon =20;
Rgoff=10;
IGon= (Udr - Uplt)/Rgon;
IGoff= -(Uplt/Rgoff);
Udrr= 4.7;  % diode reverse voltage
Irms= 2.797;
```

```matlab
Cgd1=10^10^(-12)   % Crss data sheet vds 500v
Cgd2=90*10^(-12)   % Crss Rds and Ion = 3.2V
Udd=500
Idon=1.9
Idoff=5.3
% ON switching losses
tfu1=(Udd-(Rds*Idon))*Rgon*(Cgd1/(Udr-Uplt));
tfu2=(Udd-(Rds*Idon))*Rgon*(Cgd2/(Udr-Uplt));
tfu=(tfu1+tfu2)/2
% OFF switching losses
tru1=(Udd-(Rds*Idon))*Rgoff*(Cgd1/Uplt);
tru2=(Udd-(Rds*Idon))*Rgoff*(Cgd2/Uplt);
tru=(tru1+tru2)/2;
tr=94.428*10^(-9);            %experiment values values rise
tf=47.263*10^(-9);            %experiment values values fall
%Energy ON
Eon=Udd* Idon*((Tri+tfu)/2)+Qrr*Udd
Pon=Eon*fsw;
%Energy OFF
Eoff=Udd* Idoff*((Tfi+tru)/2)
Poff=Eoff*fsw;
PswM = (Eon + Eoff )*fsw
% MOSFET losses PM + PD
PCM= Rds*(Irms.^2)*0.70
PMOSFET = PCM +PswM ;
% diodes losses
Ifrr = (2*Qrr/trr1);
EonD= Qrr*Udd *(1/4);
PonD= EonD*fsw
plot(PswM,PCM);
x=[fsw];
y=PCON;
plot(fsw,PswM,x,y,fsw,Pgd,fsw,PonD,fsw,Pdead,'--mo')
set(gca,'fontsize',30)
set(gcf,'color','white')
grid on
grid minor
ylabel('Watt');
xlabel('Frequecny');
legend({'Switching-losses','Conduction-losses','Gate-drive-
losses','Body-diode-losses','Dead-time-losses'});
grid on
grid minor
horizontal
end
```

Appendix B: PWM generation in prototype converter

```c
int main ( void )
#define PWMClockFreq                                    23E6
#define PWMSwitchingFreq80E3// this shpuld be double  the converter switching
frequency
#define PWMPeriod
(int)(((PWMClockFreq/PWMSwitchingFreq) - 1))//this the period of one PWM for
pushpull
#define Deadtime                                        50
#define PWMDuty
(int)((PWMPeriod-P1DTCON1bits.DTA))//this the duty cycle of the PWM which is
fixed to 0.5*converter period
P1TPER = PWMPeriod;
P1DC1 = PWMDuty ;
P1DC2 = PWMDuty                                                        ;
/*Initialization of the PWM for the left leg*
PWMCON1bits.PEN1H = 1;                   // PWM module controls PWMxH pin
PWMCON1bits.PEN1L = 1;                   // PWM module controls PWMxL pin
PWMCON1bits.PMOD1 = 0;
PWMCON1bits.PMOD2 = 0;
PWMCON1bits.PMOD3 = 0;//
PWMCON1bits.PEN1L=1;
PWMCON1bits.PEN1H=1;
P1TCONbits.PTEN=1;
P1DTCON1bits.DTA=55;
while(1);
}
/*------------------------------------------------------------
Function Name: InitTimer1
Description:   Initialize Timer1 for 1 second intervals
Inputs:       None
Returns:      None
-----------------------------------------------------------*/
void init_Timer1(void)
{
T1CON = 0x0000;
PR1 = 780;
/* flag */
Counter++;
}
```

Appendix C: PCB Schematics for prototype

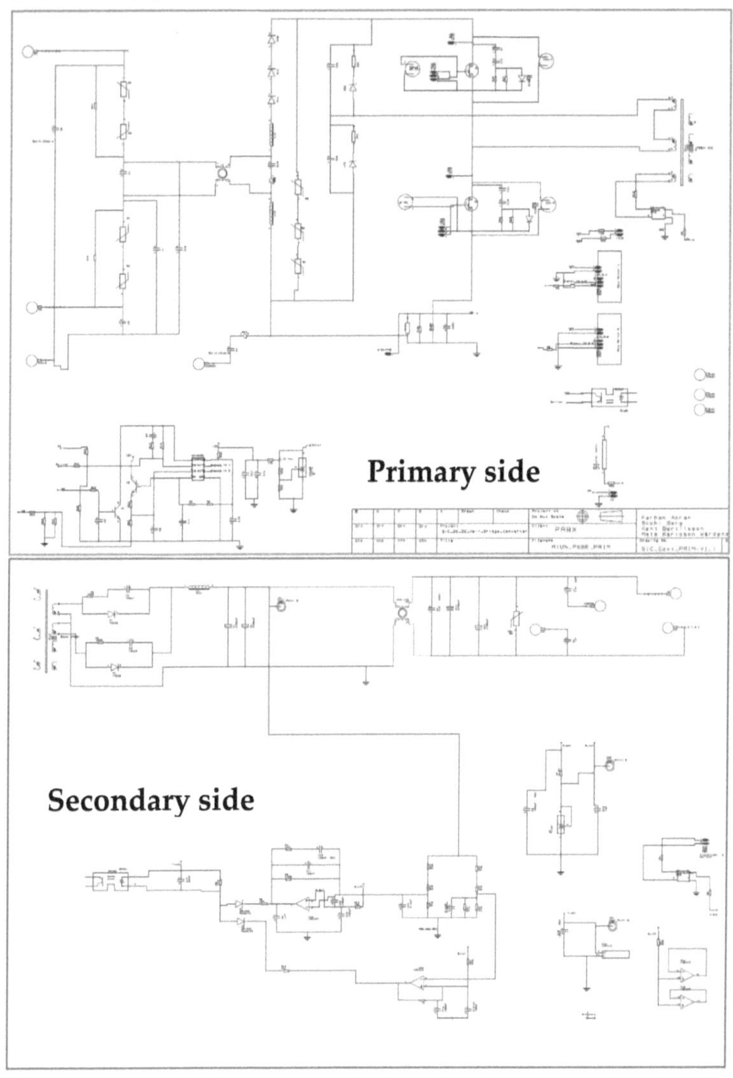

www.ingramcontent.com/pod-product-compliance
Lightning Source LLC
LaVergne TN
LVHW041715190726
843493LV00007B/2094